One Planet, Many Worlds

The MANDEL LECTURES *in the* HUMANITIES
at BRANDEIS UNIVERSITY
Sponsored by the Jack, Joseph, and Morton Mandel Foundation
Director, Professor Ulka Anjaria

The Mandel Lectures in the Humanities were launched in the fall of 2011 to promote humanistic inquiry at Brandeis University, following the 2010 opening of the new Mandel Center for the Humanities. The lectures bring to the Mandel Center each year an influential scholar or scholar-practitioner who gives a series of lectures on topics of broad interest for a range of campus audiences. The Mandel Lectures are unique in their celebration of cutting-edge topics, forms, and modes of inquiry in the arts, humanities, and humanistic social sciences: the speakers have ranged from historians and literary critics to performance artists, writers, and anthropologists. The published series of books reflects the interdisciplinary mission of the center and the wide range of extraordinary work being done in the humanities today.

For a complete list of books that are available in the series,
visit https://brandeisuniversitypress.com/series/mandel-lectures.

One Planet, Many Worlds

The Climate Parallax

DIPESH CHAKRABARTY

Brandeis University Press

WALTHAM, MASSACHUSETTS

Brandeis University Press
© 2023 Dipesh Chakrabarty
All rights reserved
Manufactured in the United States of America
Composed in Source Serif and Acumin

For permission to reproduce any of the material
in this book, contact Brandeis University Press,
415 South Street, Waltham, MA 02453,
or visit brandeisuniversitypress.com

Library of Congress Cataloging-in-Publication Data
available at https://catalog.loc.gov/
cloth   ISBN 978-1-68458-158-0
paper   ISBN 978-1-68458-157-3
e-book   ISBN 978-1-68458-159-7

5 4 3 2 1

This book was printed by Versa Press
using Natural Offset Recycle text stock with
30% post-consumer waste (PCW) content.

Bruno Latour
in memoriam

François Hartog and Étienne Balibar
in friendship and with admiration

Contents

Preface

I had the great honor of being invited to deliver the Fifth Annual Mandel lectures in the Humanities—three in all—at Brandeis University in March 2017. I had a very enjoyable and stimulating time delivering these lectures and discussing them with Professor Ramie Targoff, my kind and generous host, and her gifted and engaged colleagues who attended them. The publication of these lectures was delayed, however, through some unfortunate circumstances beyond my control, and I ended up incorporating some of the material I had presented in those lectures into my book *The Climate of History in a Planetary Age*, which was published in March 2021. This fact is acknowledged in that book.

What I present here is new material in the form of three chapters that I imagine I would have presented if I were to give those lectures today. This material is taken from research and writing I have been engaged in since the completion of *The Climate of History*, but it is material that still speaks (as does that book) to the continuing task of exploring the historical and cultural meanings of human-induced climate change from planetary, global, postcolonial—and now decolonial—perspectives. Different though this book is from *The Climate of History*, it may be read as both a prequel and a sequel to its predecessor. I have used this opportunity to explore new questions, present new material, answer some criticism, and deal with some questions I had not had time to explore before in any depth.

The experience of the pandemic—not unrelated, I argue, to the phenomenon of anthropogenic global warming or climate change—has clearly given rise to some new questions and perspectives on issues to do with the human and the nonhuman. I open with a discussion of the pandemic. I then move on to a problem that I had only touched upon in *The Climate of History*: investigating the modern origins of the separation between "natural" and "human" histories, and what may be at stake in that separation. That problem and its discussion constitute chapter 2. The third and final chapter deals with the question suggested by the title of this book: Does having different worlds make it difficult for humans to deal with a planet that is one? Many will, of course, question the proposition that the planet is indeed one, my friends Christophe Bonneuil and the late Bruno Latour among them. Latour is present in all three chapters. His thoughts have been an indispensable companion even in moments when I have not agreed with him. Philosopher Déborah Danowski and anthropologist Eduardo Viveiros de Castro are also among my main interlocuters in the third chapter. Engaging critically their efforts to decolonize anthropology, I bring their and others' thoughts to bear on what I see as the major tension running through the entire gamut of politics around climate change: our *one* planet and the many different worlds that we humans—entangled with a variety of nonhuman entities both living and nonliving—make and inhabit. These worlds themselves, I argue, are also entangled with one another.

There is one difference between *The Climate of History* and *One Planet, Many Worlds* I should mention. *The Climate of History* represented my attempt to work toward a new philosophical anthropology and remained, by choice, in a space of the pre-political: "pre-political" in the sense in which Karl Jaspers used this expression, as I explain in the last chapter of that book. *One Planet* tries to understand the main problem that haunts the calendar of (in)action of climate politics: humans are politically not-one, while Earth system scientists see the planet—the Earth system, that is—as one. It is this problem of the One and the Many that this book addresses.

DIPESH CHAKRABARTY
Chicago
October 10, 2022

One Planet, Many Worlds

Introduction
The Planet and the Political

The original title I had imagined for this volume was "Provincializing Europe in a Warming World." About five years before I gave the Mandel Lectures in 2017 on that theme at Brandeis University, I published an essay titled "Postcolonial Studies and the Challenge of Climate Change" in the journal *New Literary History*.[1] That essay in turn followed up on a remark I had made in my very first essay on climate change, "The Climate of History: Four Theses" (2009), to the effect that while I had found my training in subaltern studies, Marxist analytics of capital, and postcolonial studies adequate for discussing globalization, it fell short of the mark when I tried to think about the predicament that global warming represented for humans.[2] What I had in mind, of course, was the fact that the problem of anthropogenic climate change required us to engage with Earth System Science (ESS), the interdisciplinary branch of scientific knowledge that defined the problem of *planetary* climate change. Minus this engagement, you could still have other explanations of the erratic weather patterns we experience these days, but you would have no understanding of how the climate system of the planet, *taken as a whole*, works.

But what did it mean for humanists to engage with ESS? Did it mean that we simply added it to our repertoire of existing analytical tools in the social and human sciences without those tools being affected in any way? Would not the encounter with deep history—geobiological deep time is, after all, intrinsic to

ESS thinking—also change or challenge the shallow time of the "history" that humanist scholars work with? While I ask these questions of history as a discipline, and do so as a historian, the geographer Nigel Clark and the sociologist Bronislaw Szerszynski have already raised this question with regard to "social thought" in general, a project they describe as "geologizing the social." They argue that we should "socialize the Anthropocene," and similarly, we should "geologize" the social.[3]

My initial remark—and especially my use of the biological category "species" (as used, for instance, by E. O. Wilson)—elicited a multitude of reactions, ranging from some very hostile ones from scholars who thought all human/social/planetary problems had their origins in social and economic inequalities, to more nuanced, welcoming, and generous—but sometimes nevertheless critical—comments from scholars such as Ian Baucom.[4] And a question I was often asked at lectures and conferences was about the relationship between my work on provincializing Europe and the new work on climate change. It kept me thinking about possible connections between the two.

These criticisms and questions shaped what I said in my lectures at Brandeis in 2017. They shape the discussion in this book too. However, as I worked through the problems I had set myself in comprehending global warming and the Anthropocene hypothesis—the idea that the planet had crossed the threshold of the geological epoch of the Holocene, thanks to human impact on the planet, and entered a new epoch for which the name Anthropocene was proposed—I ended up positing an analytical distinction between the "globe" and the "planet" as categories with distinctly different imports for humanist thought. I argued that this distinction issued from my encounter with the deep time of the Earth system—the way geology and biology interacted in the history of our planet to create a life-supporting system that is now under threat. This meant, I said (helped along by some verbal and written comments from the philosopher Catherine Malabou), that the word "globe" in the expression "globalization" and the "globe" of "global warming" did not have the same meaning. The "globe" of "globalization" was at the most a 500-year-old entity brought into being by humans and their technologies of transport and communication. It was a human-told story with humans at its center. The "globe" of "global warming" referred to what ESS calls the Earth system. This was a heuristic construct of ESS scholars, which indexed planetary processes whereby geological and biological factors combined on this planet, developing thereby a system that supported complex, multicellular life. The story of the Earth system is also a human-told story, but it does not have humans at its center. Disciplines like geology and (evolutionary) biology cannot be anthropocentric, for humans come too late in the stories they tell to be at the center of their narratives. This "Earth system" of "Earth system science" is what I eventually designated "the planet."[5] The globe/planet distinction, I argued, could be foundational for a possible and new philosophical anthropology at this moment of a world-wide environmental crisis.

Furthermore, I argued that it was the intensification of the capitalist and technology-driven process of globalization in the late twentieth century that brought the domain of the planetary within the ambit of humanist thought and everyday news. This Earth system had hitherto been of interest only to specialist scientists. I wrote: "The intensification of globalization and the consequent crises of global warming, along with all the debates that have attended the studies of these phenomena, have ensured that the planet—or more properly... the Earth system—has swum into our ken even across the intellectual horizons of scholars in the humanities."[6] And then again: "Using Heidegger's language, we can say that the harder we work the earth in our increasing quest of profit and power, the more we encounter the planet. *Planet* emerged from the project of globalization. . . ."[7]

To avoid a common misunderstanding of my work, however, I should point out that I did not ever suggest that the globe and the planet constituted anything like a mutually exclusive binary. Some inattentive readers have attributed that view to me, but it is not what I said. In my thinking, "the globe" and "the planet" have always been related entities. I may refer the reader to my book *The Climate of History in a Planetary Age* for an extended elaboration of this analytical—not binary—distinction as a way of reconceptualizing human history: "For all their differences, thinking globally and thinking in a planetary mode are not either/or questions for humans."[8]

I found the globe/planet distinction useful in that it provided two different but related vantage points—the globe and the planet—from which to develop, simultaneously, two different perspectives on human history. We need to work with both of these in writing humanist histories for our age. Humans are and will remain divided on the question of how to relate to what I have called the planet. But in the age of anthropogenic climate change, the planet has emerged as an inescapable or unavoidable matter of concern. I list below some of the major differences that distinguish the "Earth system" (or the planet, in my terms) of "Earth system science" from the "globe" of "globalization":

1. The globe is human-made; it involves the work of empires, capitalism, and technology. Humans are at the center of its story; they are the main protagonist of the story of the making of the globe. The planet—also a human conceptual construct—decenters the human. Humans come too late in the geological and biological histories of the planet to be at the center of these

narratives. The category "planet" has built into it the understanding, for example, that the planet would have been 4.5 billion years old even if there were no humans to count its age.

2. The global belongs to recorded history of only the past 500 years. The planetary is about deep history, the geobiological history of the planet.

3. The globe is uniquely and singularly human, focused on human experience of the earth. The planet is comparative, emerging out of human attempts to answer questions such as, "Can Mars be made habitable for complex life?" "Did Venus become hot because it experienced runaway planetary warming?" Technology, however, connects the global with the planetary, especially when we ask, "Can the earth's climate be engineered by humans?" That question itself suggests that we are no longer simply in a global age; we are in an age that is simultaneously global and planetary.

4. Sustainability is a global and humanocentric term. It asks if humans could leave the earth in a sustainable state for humans who come after them. The planetary is about habitability, an issue that is raised, for instance, when we ask, How does a planet become habitable for life? Here "life" does not refer exclusively to life in human form.

5. Global histories are about a human-dominant order of life on this planet. The geobiological history of the planet, on the other hand, makes us realize that we are a minority form of life and that the majority forms of life on the planet are microbial— bacteria, viruses, protists, some fungi, etc. In human terms, this realization calls on us to develop minoritarian forms of thinking with regards to other forms of life.

6. The globe, the earth, and the world, as used in modern historiography, are all categorizations that assume a relationship of mutuality between humans and their worldly environment. We express this in such statements as, "The earth is our home; it is made so that we can dwell on it." The sentiment is admittedly old but it persists in new forms in modernity. The planet, i.e. the Earth system, is different in its relationship to us. We are not the end that the planetary processes supporting life are required to serve. Humans are a product of contingencies in the history of life on this planet. The planet does not return our gaze in that we cannot assume any relationship of mutuality with it.

7. The globe, forged by human institutions and technology, lends itself to moral and therefore political questions. It is amenable to questions of fairness and norms. Planetary forces, on the other hand, can reduce us to our creaturely lives. When we are faced with planetary "fury"—such as a tsunami, an earthquake, a firestorm (all of which could be triggered by our interference with the Earth system)—our politics is reduced to the politics of survival, something that Kant or Arendt would not call "politics" for it is bereft of any sense of morality.[9]

The Planet and the Political

The category "planet"—as I have used and explained it here—poses a peculiar challenge to what humans regard as the domain of the political, an area that includes a range of activities from those issuing from the nation-state and other formations of power and inequality down to the individual human. The problem is this: the planet, that is the Earth system, is both differentiated and unitary. It is, as I have called it, a "dynamic ensemble of relationships."[10] It lurches from one state to another over geological time.[11] Clark and Szerszynski, who have done more than most to integrate this idea of the planet with the social sciences, observe that "the Earth . . . is more than a single system that supports life, . . . there are decouplings or disconnections in the planetary body, as well as couplings."[12] They introduce the idea of "planetary multiplicities," by which they refer to the undeniable fact that "the Earth has an inherent potential to shift from one state to another and to do this quickly." But they also acknowledge the oneness of this planet and see it as "a dynamic and self-organized" entity, as do Mark Williams and Jan Zalasiewicz in their recent book on the history of the biosphere.[13]

In its differentiated representation, the planet or the Earth system speaks of various entanglements—of the geological with the biological, of different forms of life with one another, of connecting the work of bacteria and planktons, for instance, to the proportion of oxygen in the air. Such entanglements are subjects on which Donna Haraway, Anne Lowenhaupt Tsing, Bruno Latour, and others have written with insight and imagination. Entanglements represent the differentiated aspect of the planet. Yet the planet is also one. The "carbon budgets" for humanity that the Intergovernmental Panel on Climate Change (IPCC) publishes from time to time are based on the idea that there is one atmosphere and one planet, a planetary climate system that can be treated as a whole and a single Earth system supporting life.[14] It is significant that the "science" of all this is called "Earth system" (and not "systems") science. The definition provided by the now-closed International Biosphere-Geosphere program run by Earth system scientists is still relevant:

> The term "Earth system" refers to the earth's interacting physical, chemical, and biological processes. The system consists of the land, oceans, atmosphere and poles. It includes the planet's natural cycles—the carbon, water, nitrogen, phosphorus, sulfur and other cycles—and deep Earth processes. Life too is an integral part of the Earth system. Life affects the carbon, nitrogen, water, oxygen and many other cycles and processes. The Earth system now includes human society. Our social and

economic systems are now embedded within the Earth system. In many cases, the human systems are now the main drivers of change in the Earth system.[15]

All this is not to deny that there could be many other understandings and representations of the planet than that proposed by ESS. Indian astrologers, for instance, could have an entirely different set of reasons explaining the planetary plight of humans. From the prehistoric humans who settled the islands of the Pacific thousands of years ago, navigating the seas by the night sky, to ancient Greek and Indian astrology and peasant sayings about seasons, through to the Copernican revolution in the sciences and its consequences—these are all instances of planetary thinking. But the IPCC recommendations regarding carbon budgets for the atmosphere of the planet could only be made on the assumption that *the planet is differentiated and yet one*.

Humans, on the other hand, are *only* differentiated, i.e., not-one, politically speaking. There is no one humanity in politics. Politics stems from what differentiates humans. That which may be a universal truth about humans—that they are a biological species entangled with other forms of life, for instance—constitutes no ground from which a unitary political subject could arise and project itself on the world. As Sartre, following a train of thought that went back to Hegel, once pointed out in his preface to Frantz Fanon's *The Wretched of the Earth*, the recognition of one human by another as a human being may very well be the ground for the production of hierarchy and separation, colonization, torture, oppression, dispossession, and enslavement: "We [Europeans] saw in the human species an abstract premise of universality that served as a pretext for concealing more concrete practices: there was a race of subhumans overseas who, thanks to us, might, in a thousand years perhaps, attain our status. In short, we took the human race to mean elite."[16]

The IPCC may posit one planet and issue calls for somewhat synchronized human action based on a unitary planetary calendar, but humans and their representative institutions would always want to route this oneness through their differences. The one planet of ESS splinters into the many planets of the rich and the poor, of the formerly imperial nations and of the peoples they once subjugated or still do. A very powerful and strong articulation of this point of view may be found in Kathryn Yusoff's *A Billion Black Anthropocenes or None*:

> If the imagination of planetary peril coerces an ideal of "we," it only does so when the entrappings of late liberalism become threatened. This "we" negates all responsibility for how the wealth of that geology was built off the subtending strata of indigenous genocide and erasure, slavery and carceral labor, and evades what that accumulation of wealth still makes possible in the present—lest "we" forget that the economies of geology still largely regulate geopolitics and modes of naturalizing, formalizing, and operationalizing dispossession and ongoing settler colonialism.[17]

And later:

> The championing of the collective in geology under the guise of universality or humanity is actually a deformation of the differentiation of subjective relations made in and through geology. This is how the codification of geology (as land, mineral, metal, gold, commodity, value, resource) becomes the historical basis of theft, actioning a field of dispossession in which the language of containment is used to materially organize extraction, where violence is covered in the guise of liberating surplus wealth from people and the earth.[18]

Yusoff has the histories of Indigenous dispossession and "slavery and carceral labor" in sight in making these remarks. Our knowledge about the Earth system rests on a historical and ever-present foundation of racist oppression. The "one-ness" of the Earth system hides the fact of differentiation of humans that is the condition of its own possibility. Even when we turn to nations like India and China, nations that did not suffer settler-colonial rule but were victims of European empires in different ways, we see the same tendency, in practice, of splitting the planet. When China and India say to the industrialized nations that they, the former, should be allowed some carbon space for development as warming of the planet is the historical responsibility of the latter, they not only raise an important question of climate justice, but they also, by the same move, differentiate the planet. It is like saying to the developed nations, "The planet is mainly yours for the next two or three decades. Looking after it is also mainly your business. We need to focus predominantly on growth for the time being." That is an act of splitting the planet.

This structural and unresolvable mismatch between the oneness of the Earth system as imagined by the science of climate change and the pluriversal quality of human politics defines, to my mind, a fundamental aspect of the human condition today. ESS itself is a product of the differentiated nature of humanity; it is a product of the divisions of the Cold War. The Cold War profoundly shaped atmospheric science and comparative planetary science: nuclear fallout affecting the atmosphere, weaponization of weather, and colonizing Mars and other heavenly bodies became security concerns of the superpower nations. The technology of space exploration, after all, came out of the Cold War and contributed to the growing weaponization of atmosphere and space. All that eventually created ESS in the 1980s in the United States. The connection may even be seen in the fact that climate scientists would not have been able to bore into ice created eight hundred thousand years ago and capture ancient air bubbles if the US defense establishment and the much-denounced oil and mining companies had not developed the necessary technology for drilling, which was then modified to deal with ice. Whether we are in the Anthropocene or not, the scientific explorations of the condition of global warming allow scientists to construct—through use of satellite measurements, drilling into the polar ice caps, oceanic measurements, etc.— a hyper-object called the "Earth system." It is this that I am calling the planet. It is an abstract scientific construct that is meant to explain how this planet, where geology and biology have long been connected phenomena, has a life-supporting system that works as just that, a "system." Planetary processes—including the work done by phytoplankton, for instance—have sustained a level of atmospheric oxygen that is helpful for the survival of animals (including humans) and plants. This atmosphere is critical for our existence, but it was not made with humans in view. The planet has maintained this atmosphere for nearly 375 million years.[19]

To place the origins of ESS in the context of the Cold War is to historicize the science, but it does not detract from the knowledge the science has come to embody. Widespread awareness of this planetary condition of human politics could only have come about as the increasing frequency and impact of extreme weather events across the world sensitized us to the messages the IPCC had been broadcasting for a while.

My book *Provincializing Europe: Postcolonial Thought and Historical Differences* spoke to the problem of the global, but it was unaware—as were most postcolonial critics and theorists of the time—of the planetary.[20] As I write this sentence, I am reminded of Baucom's thoughtful, attentive, gracious, and yet sharp critique of my climate-related writings in his book *History 4° Celsius: Search for a Method in the Age of the Anthropocene*, where he argues, rescuing me from myself as it were, that there are indeed resources in

Provincializing Europe that may still be of use in conceptualizing the geological agency of humans.[21] He may or may not be right, or there may be some middle ground between our positions—let me for now reserve judgment on that question. But I would still submit that the problem of the planet as I think of it—that is, the planet as the Earth system—does not emerge in Baucom's book, dependent as it partially is on Paul Gilroy's use of the word in the expression "planetary humanism," where the "planet," in my reading, remains coextensive with humanity.[22] My use of the word "planet" refers to the Earth system, something that has been in existence for hundreds of millions of years, from long, long before humans even came into being. Unlike in many other forms of planetary thinking, this planet does not index what humans can experience directly about the world. We have no immediate experience, for example, of the multifarious roles the deep seas, oceanic currents, the trace gases in the atmosphere, the Ozone layer overhead, the Siberian permafrost, or the Himalayan glaciers may have in maintaining the climate system of the planet as a whole. It is also impossible to collapse the entire history of the planet into the history of human existence. Some of the critical planetary processes work on scales of time much bigger than what humans usually consider in making political and other decisions. What I mean by the planetary is perhaps more in congruence with Achille Mbembe's statement that "humans are a part of a very long, deep history that is not simply theirs; that history is vastly older than the very existence of human race, which, in fact, is very recent."[23]

However, the political question, What is to be done in a climate emergency? seems to be for humans alone. Other living beings will no doubt have to respond to the challenges of a warmer planet, but the articulated question is for humans alone. It is a question that is sharpened for humans as extreme weather events take place in different parts of the world and as planetary environmental crises in different forms—such as the pandemic—imperil human and other lives. However abstract the conception of the Earth system is and however entangled the human may be with other forms of the living and the non-living, it seems to me that there is, as of now, no alternative to beginning with the question of human experience of the world in conceptualizing the political. Human phenomenology has to be granted (in old Althusserian terms) a "relative autonomy" with respect to histories of nonhumans including the history of humans as a planetary thing. (This is not to deny the agency of the nonhuman—including the human as a thing-like geophysical force—in shaping humans and their experiences.)

I may explain this point with reference to Thomas Nail's original and admirable book, *Theory of the Earth*.[24] I disagree, respectfully, in some fundamental ways with what Thomas Nail does, which is to collapse human history completely into the history of the planet in a well-intentioned but misconceived attempt to avoid anthropocentrism. Because humans are a product of the earth—many have pointed to the etymological connection between "human" and "humus"—their history, Nail argues, fundamentally repeats features of the history of this living planet. He asks, "Where did the human idea of respiration itself come from in the first place? Before there were breathing humans, . . . there was already a breathing earth. Humans breathe because the earth breathes. . . . The atmosphere is the breath of the earth. This is not a metaphor—or perhaps it is an inverted one. The earth does not breathe because the humans breathe, but the other way around. Anthropocentrism is already a *terramorphism* that has forgotten itself."[25]

Not humans but the earth, the planet from even before life appeared on it, is truly the subject of the history that Nail writes: "The earth knows itself *through* the human body (and other bodies) not because of any logical or ontological necessity but because of the practical, historical, and material way that the earth has come to iterate itself, billions of years before their existence."[26] Sometimes Nail scales the subject up to cosmic proportions: "Human animals do not live *on* the earth. They *are* the earth. Technically, even the earth itself is not *of* the earth. It is a branch of a massive cosmic dendritic flow, of which humans are one tiny experimental capillary seeking out optimal kinetic expenditure into the cool darkness of space."[27]

Yet Nail asks of humans that they "help our planet and our cosmos expend itself": "The duration of our life on this planet is related to how well we help our planet and the cosmos expend itself."[28] Historical characters like "capitalists," "capitalism," "anthropocentrism," and even a human "we" populate—however sparsely—some of Nail's pages: "Capitalism and anthropocentrism are destroying the planet, and *we* ought to stop them if *we* want to survive."[29] But how would this "we" or even "capitalism" exist as a category if, conceptually, we did not grant humans and their institutions some degree of autonomy vis-à-vis the cosmic history of the planet even if we accepted the inseparability, in the ultimate instance, of these two histories?

This autonomy does not mean that we have solved the problem of the "we"—the collective subject—of human history. The problem of the "we" is, in fact, the most critical aspect of our current planetary crisis. There is no one "we" to respond to a planet or an Earth system that is, by the contingency of Earth's pasts, one. If the evidence of human history is anything to go by, there never has been a one "we" of humans. This is why some of the concerns of *Provincializing Europe* are still relevant today when humans find themselves situated on the cusp of the global and the planetary.[30]

The three essays in this book all focus on problems that have arisen in recent times out of this jagged and mismatched interface between the "global" and the "planetary."

The first essay thinks through the current pandemic, treating it as an event inseparable from the planetary environmental problems we face today. I see the pandemic as representing an entanglement of the time of human history—the period of the Great Acceleration when nations in Asia, Africa, and Latin America join the race to industrialize and consume—with the time of a more Darwinian history of the evolution of bacteria and viruses.

The second essay considers some of the problems that a conception of humans as a very large geophysical force impacting the planet, like an asteroid might, poses for postcolonial historiography of modernity. That historiography has been focused since the 1960s on different emancipatory philosophies of human freedoms, mainly from oppression and poverty. Both in this essay and the essay that follows, I underline the genealogy and the continuing political importance of these postcolonial visions of "development as freedom" (to use Amartya Sen's famous expression).

The third and concluding essay of this volume goes to the heart of what I have already referred to as "the One and the Many" problem that makes climate change such a difficult issue to tackle. To repeat: ESS posits one Earth "system"; they use the word in the singular. There is, however, no corresponding single "humanity" that is either responsible for the warming or can act as one in combating it. This mismatch produces a sense of disorienting times that François Hartog, Latour and others have commented on.[31] Yet the IPCC's talk about there being a "climate emergency" clearly has a temporal dimension.[32] Emergency connotes a sense of time when urgent action is needed. But the time in question is planetary time, the time it will take to remove excess carbon from the atmosphere, and it is based on the assumption that the planet—with its atmosphere and the seas—is one. But this oneness of the Earth scientists' planet is vigorously disputed from within European thought and postcolonial and decolonial criticisms. Many scholars seeking to "decolonize" knowledge point to the unreconcilable differences between these "modern" and Indigenous thought, for example. The last chapter in this book argues, however, that in an increasingly global world, these otherwise distinctive intellectual traditions are already always entangled with one other. Any human sense of planetary emergency will have to negotiate the histories of those conflicted and entangled multiplicities.

Human differences are, in the end, existential. Just a fact of life. We are born different from one another, a fact in which Hannah Arendt saw grounds for hope. Similarly, Aristotle's deliberations on justice, equality, and proportionality in *The Nichomachean Ethics* focused on the problem of difference and exchange. Exchange was central to the question of forming a community that was always made up of people who were necessarily "different and unequal ... for it is not two doctors who associate for exchange but a doctor and a farmer, or in general people who are different and unequal; but these must be equated ... must be somehow [made] commensurable."[33] On the ground, then, there is only difference. The task of politics is to find solidarities across these differences, sublating, articulating, or even suspending them for a while. But how long is "a while," and how long does it take to get there? Issues of temporality are a fundamental aspect of human politics. Politics is framed by a sense of community that—to speak with Derrida—is always to come. This essential feature of politics now itself emerges as a political problem if some humans declare "a climate emergency." Should humans patiently work through their differences even if they miss the schedule of a planetary calendar of action that the IPCC recommends? Or should powerful nations bypass the difficult politics of difference and launch into climate engineering? Or should humans, finding themselves entangled in other forms of life, "stay" actively, not passively, "with the trouble" (as Haraway puts it) or "with the present" (as I put it, with grateful acknowledgment of Haraway's text) and work toward recomposing the commons?[34] But what would the political commons between microbes, constituting the majority forms of life, and humans, a dominant minority, look like?

I do not have a grand solution to these questions though they are some of the larger and urgent political questions of our time. But I do suggest through the three studies assembled here that the political may indeed be something that—as of now—remains parochially and provincially human. The facts of human entanglement in other forms of life are visible as much at the planetary level—in the history of the Earth system, for instance—as in the story of the current pandemic that affects us collectively and personally. But these entanglements—although real—do not constitute in themselves political subjects while it is possible for humans to act as advocates (within human institutions) for certain other forms of life such as animals, plants, or even non-living entities like mountains or rivers. Faced with the cascade of events or problems that we call anthropogenic climate change, both humans and nonhumans will act. Trees, animals, birds, fish, and other creatures and microbes will all seek to move, for instance, to sites and areas more congenial to their flourishing. But the proactive question "What is to be done?" is still for humans alone. Grasping the analytical distinction between "the globe" and "the planet" may give us a handle on this historical experience of disorientation.

If someone were to ask me today, "What do I provincialize in my climate-related work?" I would probably say that I provincialize the human and do so as a humanist historian of human beings.[35]

1
The Pandemic
and Our Sense of Time

The pandemic and the climate crisis are connected phenomena. One could say that they both speak of Anthropocene times. The story of rapid global economic growth—the history of capitalism in all its different varieties (imperial, liberal, and neoliberal)—is common to narratives that underpin discussions on both crises. They both arise from what has been called the period of Great Acceleration in global history, the expansion—or more aptly the explosion—of the human realm over the twentieth and the twenty-first centuries, beginning especially in the 1950s. In an increasingly extractive relationship to the earth, this expansion claimed more and more of the products of the planet's biosphere, i.e., from the part of this planet that immediately sustains life. Bruno Latour and others, following scientists like Timothy Lenton, have called this "the critical zone" of the earth.[1] The key to this expansion, as we all know now, was the cheap and plentiful energy extracted first from coal and then later from oil and gas, all of them different kinds of fossil fuels. More than 87 percent of the total consumption of fossil fuels by humans and their institutions has taken place in the period from reconstruction of the industrialized economies after the Second World War to the present. This is why the Great Acceleration is dated by historians and Earth system scientists from c. 1950.[2]

The twentieth century became "a time of extraordinary

change" in human history: "The human population increased from 1.5 to 6 billion [roughly fourfold], the world's economy increased fifteenfold, energy use increased from thirteen- to fourteenfold, freshwater use increased ninefold, and the irrigated areas by fivefold."[3] To add some more dramatic figures, the world's urban population increased in the same century by 1,200 percent; industrial output by 3,500 percent; energy use by 1,200 percent; oil production by 30,000 percent; water use by 900 percent; fertilizer use by over 30,000 percent; fish catch by 6,500 percent; organic chemical production by 100,000 percent; car ownership by a staggering 775,000 percent; and the carbon dioxide in the atmosphere rose by 30 percent.[4] The very well-known Great Acceleration graphs produced by Will Steffen and others show that, for most of these figures, the growth became exponential around 1950, rising even more steeply in the 1980s when China and India liberalized their economies and joined the race for industrialization and modern consumption with greater efforts.[5]

In addition, the German scholar Hannes Bergthaller has used a 2017 survey by the Brookings Institution to offer some telling statistics demonstrating an acceleration of the human consumption of resources as well. It was "only around 1985," reports Bergthaller, "that the [global] middle class reached 1 billion people, about 150 years after the start of the Industrial Revolution in Europe," but it only took "21 years, until 2006, for the middle class to add a second billion," much of this reflecting the extraordinary growth within China. Bergthaller continues, "The third billion was added to the global middle class in nine years. Today we are on pace to add another billion in seven years and a fifth billion ... by 2028."[6] No wonder that humans also emerged in this period as the biggest geomorphological agent on the earth, shaping its landscape and the continental shelves in the oceans, and as a geological force changing the climate system of the whole planet, ushering in, as some scientists suggest, the new geological epoch of the Anthropocene.[7]

"The principal reason," writes Bergthaller in his essay on Asia and the Anthropocene, "why all the curves of the 'Great Acceleration' are still pointing relentlessly upwards (with the notable exception of that for population . . .) is the spread of middle class consumption patterns around the world, if by middle class we understand people with a household income sufficient to purchase consumer durables (such as refrigerators, washing machines or motorcycles), to spend money on entertainment and on the occasional vacation." As recently as 2000, he adds, "about 80% of this 'global middle class' was living in Europe and North America. . . ." But by 2015, "their share had dropped to about 35%, due largely to the rapid expansion of the middle class in Asia." Bergthaller reports that by 2030, "the Asian middle class" is expected to be "at least three times larger than that of the old 'West,'" and will account "for two thirds of the world's total. . . ."[8]

The Anthropocene thus produces a peculiar sense of historical time, something we could call its "chronopolitics." I owe this word to three younger scholars—Tobias Becker, Christina Brauner, and Fernando Esposito—who organized an online conference of this title on 16–18 September 2021 and glossed it to mean "[the] time of politics, politics of time, politicized time."[9] I, however, mean something slightly different. Because of the multiple ways in which the planetary environmental crises we gather under the name Anthropocene play out on different scales of time and space, both human and nonhuman, the Anthropocene, it seems to me, fragments human futures in unprecedented ways. One could, for instance, tell the story of the Anthropocene as that of a crisis of neoliberal capitalism, a crisis of the industrial and consumption-oriented ways of human life, as a crisis of biodiversity leading to a sixth Great Extinction of species, or as a story of how humans fended off the next ice age by many, many thousands of years. These futures do not all happen on the same scales of time and space. The Anthropocene itself, being a geological epoch, may last much longer than humans—a point that raises a question about whether it could at all be used as a periodizing device for human history. But the Anthropocene also produces very short-term futures for humans—so short-term that one could think of them as "the present." Our sense of the time of the pandemic contains particular and entwined figures of the historical present and the historical future. Much talk about post-pandemic futures is in nature nostalgic, expressing a desire to return to the ease and comfort of the pre-pandemic times; but the politics of and the demand for "equal access to vaccination" convert this future time into a present that we want to fully—and equally—inhabit.[10] What I explore in this chapter is the figure of the pandemic as a time of the present, one that makes the future hard to imagine.[11]

The Pandemic and the Great Acceleration of Human History

We are now being told by numerous infectious diseases specialists that we live in an era of pandemics. Pandemics and epidemics have accompanied humans ever since the invention of agriculture and the domestication of animals. Hunter-gatherer communities also suffered infectious diseases but, as some virologists put it, "like the sparse populations of our primate relatives, they suffered infectious diseases with characteristics permitting them to persist in small populations, unlike crowd epidemic diseases." They go on to say that agriculture with the concomitant domestication of animals played "multiple roles in the evolution of animal pathogens into human pathogens."[12] It took humans thousands of years to strike an equilibrium with these zoonotic diseases. But the difference today is this: These crises of the past "were once separated by centuries, or at least many decades," write the infectious diseases specialist David Morens and his coauthors in a recent paper (2020), but the emergence of new diseases is now becoming a more frequent phenomenon.[13]

Starting with the year 2003, Morens and his colleagues counted the outbreaks over the course of seventeen years with at least five pandemics or potential pandemics affecting the world: severe acute respiratory syndrome (SARS, 2003), "a near pandemic" of influenza (H1N1 pdm, 2009), a chikungunya pandemic (2014), a Zika pandemic (2015), and over 2014–15, a "pandemic-like extension of Ebola over five African countries."[14] They grant that "the meaning of the word 'pandemic' has recently been reinterpreted according to differing agendas," and yet they conclude with a sentence that sums up the risks of our times: *It is clear that we now live in an era of pandemics*, newly emerging infectious diseases, and the return of old contagious foes."[15] A more recent paper by Morens and his colleague Anthony Fauci, director of the National Institute of Allergy and Infectious Diseases in the United Sates, comes to the same conclusion:

> Newly emerging (and re-emerging) infectious diseases have been threatening humans since the neolithic revolution, 12,000 years ago, when human hunter-gatherers settled into villages to domesticate animals and cultivate crops.... Ancient ... diseases with deadly consequences include smallpox, falciparum malaria, measles, and bubonic/pneumonic plague.... [But] the past decade has witnessed unprecedented pandemic explosions: H1N1 "swine" influenza (2009), (2014), and Zika (2015), as well as pandemic-like emergence of Ebola fever over large parts of Africa (2014 to the present).... *One can conclude from this recent experience that we have entered a pandemic era....*[16]

All of the pandemics just listed—plus the Middle East respiratory syndrome (MERS) that spread to humans from dromedary camels in 2012 that is not named—are zoonotic in origin, i.e., they are infections that have resulted from viruses and bacteria switching hosts from animals to humans, sometimes via other animals, in recent times. A 2005 inquiry found that "zoonotic bugs accounted for 58 percent" of 1,407 "recognized species of human pathogen."[17] A 2012 review of the 6th International Conference on Emerging Zoonoses, held in Cancun, Mexico, on 24–27 February 2011 with eighty-four participants from eighteen countries noted that "some 75% of emerging zoonoses worldwide" had "wildlife origins." The review also found global trade in wildlife and the continuous destruction of animal habitats contributed to the problem.[18]

"Human beings are the ultimate causes of pandemics," assert Morens and his colleagues. They point out that it is "deforestation, agricultural intensification, urbanization, and ecosystem disruption" that "bring people into contact with wildlife and their potentially zoonotic pathogens."[19] "To put the matter in its starkest form," says science writer David Quammen, "human-caused ecological pressures and disruptions are bringing animal pathogens ever more into contact with human populations, while human technology and behavior are spreading those pathogens ever more widely and quickly."[20] He mentions the critical factors at work here. Humans are

> causing the disintegration . . . of natural ecosystems at a cataclysmic rate. Logging, road building, slash-and-burn agriculture, hunting and eating of wild animals, . . . clearing forest to create cattle pasture, mineral extraction, urban settlement, suburban sprawl, chemical pollution, nutrient runoff to the oceans, mining the oceans unsustainably for seafood, climate change, . . . and other "civilizing" incursions upon natural landscape—by all such means, we are tearing ecosystems apart.[21]

The United Nations Environment Programme's *Preventing the Next Pandemic: Zoonotic Diseases and How to Break the Chain of Transmission* and *The Loss of Nature and the Rise of Pandemics* published by the World Wide Fund for Nature support these conclusions.[22] They see some "major anthropogenic drivers of zoonotic disease emergence": (1) increasing demand for animal protein, particularly in Asia and in sub-Saharan Africa; (2) unsustainable agricultural intensification, in particular of domestic livestock farming that "results in large numbers of genetically similar animals" that are more vulnerable to infection (swine flu being a case in point); (3) increased use and exploitation of wildlife; (4) unsustainable use of natural resources accelerated by urbanization, land-use change, and extractive industries that include mining, oil and gas extraction, logging, etc. encouraging "new or expanded interactions between people and wildlife"; (5) the increasing amount of human travel and trade; (6) changes in food-supply chains driven by "increased demand for animal source food, new markets [including "wet" markets] for wildlife food, and poorly regulated agricultural intensification"; and (7) climate change as "many zoonoses are climate sensitive and a number of them will thrive in a warmer, wetter, and more disaster-prone world foreseen in future scenarios."[23] The conclusions drawn in the World Wide Fund report are very similar:

> Human activities are causing cataclysmic changes to our planet. The growing human population and rapid increases in consumption have led to profound changes in land cover, rivers and oceans, the climate system, biogeochemical cycles and the way ecosystems function—with major implications for our own health and well-being.... Land-use change, including deforestation and the modification of natural habitats, are responsible for nearly half of emerging zoonoses.[24]

That we did not have this tragic global pandemic a decade or so ago now appears to have been purely a matter of human luck. A team of scientists in Hong Kong warned the scientific community some thirteen years ago, in 2007, that since coronaviruses were "well known to undergo genetic recombination" that could lead to "new genotypes and outbreaks," the "presence of a large reservoir of SARS-Co-V-like viruses in horseshoe bats, together with the culture of eating exotic mammals in southern China, is a time bomb. The possibility of the reemergence of SARS and other novel viruses from animals or laboratories and therefore the need for preparedness should not be ignored."[25] The warning was not heeded.[26] Quammen reports scientists guessing since 2012 or thereabouts as to when a pandemic, the "Next Big One" with "high infectivity preceding notable symptoms," would come.[27] For, as Quammen puts it, "If you are a thriving population, living at high density but exposed to new bugs, it's just a matter of time until the Next Big One arrives."[28] But nobody was listening in either 2007 or in 2012.

The Pandemic as Presentism

One could say that the pandemic produces for us a present in which all talk of moving beyond the pandemic to a "normal" future sounds like a desire for a backward movement, to go back to what we had before. This is not the presentism, then, that François Hartog wrote about in his celebrated book, *Regimes of Historicity*, where he describes a post-war Europe experiencing the collapse of all futures into its war-weary present.[29] In the pandemic, the future arrives as nostalgia. In the pandemic, there is a present without a future that is also not at the same time about moving back to the past. A present that is ever present in that sense, not the vanishing present one usually reads about in modern discussions of the past, present, and future. It is also a present that all humans can fully inhabit—cognitively and affectively—as their "now." When we ask for a just distribution of vaccines through the world—or even when we resist vaccines on secular or religious grounds—we inhabit that present. We look at the past pandemic of 1918 to ask how long this one might last. That expected duration—a few years, four years the last time—defines this present.

But the pandemic has also registered a profound shift in the constitution of the "everyday normal" for the late-modern and urban humans of *the post-antibiotic period* in medicine—the "heirs of the industrial and imperial impetus," as Pierre Charbonnier describes us.[30] The simultaneous acknowledgment and forgetting of deep, geobiological histories of life and of the planet—of the ocean of microbes that is both inside and outside our bodies—were often contained in the phatic aspect of our everyday exchanges.[31] When we greeted each other with a remark on the weather, we acknowledged, as it were, the work of the sun, clouds, wind, trees, plants, light and shade—the planetary, in short—but only for a brief moment, before transitioning on to what Roman Jakobson called "informative communication" that was much more closely tied in our practices to the more important business of advancing our individual and collective human ends. The planetary was thus something we considered in separation from the more important business of everyday life, and treated it as something contained in the phatic, the pleasantries we use to begin a conversation.[32] I say "the late-modern and urban" human, for someone in a rural or Indigenous context, a deficit of sunshine or rain would have more immediate and palpable consequences.[33] The phatic utterance in the case of the late-modern, urban, post-antibiotic person was a measure of the cultural distance from or of the indifference they "normally" experienced with regard to the deep-historical work of all that sustains life on the planet.

A "normal" moment for us, then, is one that allows us to forget or ignore the life-altering work that microbes do even when we are not in a position, intellectually, to deny their presence. I owe this insight to some fascinating observations that historian Arvind Elangovan kindly shared with me on reading my book *The Climate of History in a Planetary Age*, published in 2021.[34] He recalled how common it was, in his experience, for letters written by twentieth-century Indians to carry news about the physical illness of the writer or the recipient even if that did not constitute the main point of the letter (Elangovan is a historian of the Indian constitution).[35] "In many of the writings that I have seen of nationalist leaders, such as Ambedkar's papers or letters written by B. N. Rau or Shiva Rao even," he wrote, "a frequent ... [part] of the letters was ... [where] they would note how sick they were or how they were recovering or ... [asked after] the health of the recipient of their letters." He added, "Indeed, in Tamil, the first sentence that my mother would always write in those (good old!) Inland letters to me or to my relatives was 'Nalam, nalam ariya aaval'—literally translated as 'Fine, yearning to hear that the same is true of yours.'"

"These moments," Elangovan wrote, "seem to me to register a cognition at the barest minimum ... an acknowledgment of the microbial, bacterial, and/or the viral (but, of course without a conscious recognition of the same, mostly). But it was just that. Immediately, that polite enquiry was succeeded by the main intent of the letter. It is as though every letter began with a parenthetical acknowledgment of the species aspect of our lives, to be quickly swept away and transitioned to the human aspect of our lives! Unless, of course, ... the person was seriously sick, ... [when] the question immediately got translated in institutional terms—to questions such as 'what did the Doctor say?' or 'what is the Hospital saying?', etc. ..."[36]

We will not get involved here in debates on whether phatic speech—first commented on by the great anthropologist Bronislaw Malinowski in 1920—signifies "communion" between humans (an overcoming of a threatening silence) or merely a matter of communication.[37] We will simply register that the pandemic connotes a time when our recognition of the microbial world we live amid cannot any longer be contained within the phatic and thus forgotten as we go about our everyday lives. The question, "How are you?" could not, in the situation in 2020–21, be a simple, conversation-starting statement. We often indicated this by making the phatic part of our communication routinely register the strangeness of our times. Almost every new email I received those days began with an expression of concern about the "strange" or "disturbing" times we were passing through. In fact, it would have been considered brusque to begin a new email message without an expression of this concern.

The fact that the offending virus can no longer be contained in the structure of the phatic has some ironical implications both for the history and the theory of biopolitics as it was enunciated by Michel Foucault in the 1970s. Let me remind you of a particular day: 8 February 1978. Foucault was already engaged in giving a series of public lectures at the College de France elaborating on his idea of bio-power and the governmentalization of the state. Everything was apparently going well until this day arrived when Foucault felt unwell as he stood—at the lectern or pulpit (as the French say), I imagine—to begin the fifth lecture of the series. He had a touch of the flu. He began with an apology: "I must apologize, because I will be more muddled than usual today. I've got the flu and don't feel very well." Yet he wanted to proceed with the lecture as he had "some misgivings" about first letting his audience gather and then telling them to leave "at the last minute." So, he decided to talk "for as long as [he could]" and asked in advance for forgiveness for both "the quantity as well as the quality" of what he had to say.[38]

Think, then, of what is happening to Foucault's categories today. Biopolitics was about securing the biological life of a population, an extension of Montesquieu's anticipation that "politics [was] really about making life last a little longer."[39] Foucault was clear that the category "population" brought the question of "nature" into politics. He began his 1978 lectures at the College de France on 11 January with this statement:

> This year I would like to begin by studying something that I have called, somewhat vaguely, bio-power. By this I mean a number of phenomena that seem to me to be quite significant, namely, the set of mechanisms through which the basic biological features of the human species became the object of political strategy, of a general strategy of power, or, in other words, how, starting from the eighteenth century, modern Western societies took on board the fundamental biological fact that human beings are a species.[40]

Reading him today, I find his use of the word "species" a little misleading, for he was not speaking of humanity constituting a biological species as such; nor was he writing a Darwinian version of evolutionary history of species in which something like natural selection would have been a determining factor.[41] That deeply natural-historical dynamic was beyond what interested Foucault. He was thinking of humans individually carrying certain evolved needs and capacities—the need to eat, the propensity to sustain life, to procreate, to age, to suffer diseases—that they owed to the fact of their being members of a biological species. Yet it was through his observations on the political development of strategies for governing the health and lives of "populations"—managing demographics—that the deeply natural-cum-biological history of the human species entered Foucault's meditations on power.

The growth of cities and problems of overcrowding leading to "more diseases" and "more deaths" were central to Foucault's formulations: "It seems to me," he wrote, "that with this technical problem posed by the town ... we see the sudden emergence of the problem of the 'naturalness' of the human species within an artificial milieu. It seems to me that this sudden emergence within the artifice of a power relation is something fundamental ... [to] what we would call biopolitics, bio-power."[42] The concern with the governance of lives meant that states had to evolve strategies to deal with crop failures, climate, and the supply of grains for the management of epidemics, diseases, famines, and mortality, all of this making "population" into a category that would never lose its "naturalness" for Foucault. It would almost acquire an autonomous, "natural," thing-like status in Foucault's understanding of the state's political calculus, something that had to be managed by a discursive institutional regime stretching well beyond the issue of political sovereignty.[43]

Foucault was very clear, though, that while the natural entered the political via the category "population," his account of the biopolitical was not a piece of natural history. After all, humans' theories of nature, he argued (mistakenly, it seems, from today's vantage point), did not affect nature: "It goes without saying that the fact that since a certain point of time we have known that the Earth is a planet has had no influence on the Earth's position in the cosmos."[44] But not so with "population" as a "reflexive prism" of the state.[45] This prism affects human-institutional practices and their object, the population. In that sense, "population" is a category like "forests," something to be managed by humans. Like "forests," "population" is a piece of nature refracted through strategies of power; it does not belong to the deep history of evolution. For Foucault, then, natural history remains, ultimately, separate from human history. As with the statement by Elangovan's mother in her letters, the virus that afflicted Foucault on the day of his fifth lecture comes to us only as a trace of something that registered its presence and yet remained unacknowledged in the phatic overtures of Foucault's prose.

What we have with the pandemic, however, is the fact that the phatic cannot contain the novel coronavirus or SARS-CoV-2 anymore. As I have noted, we cannot at present ask anyone how they are with complete indifference to the virus. The intensification of bio-power or biopolitics—the unbridled, accelerated, and extractive mobilization of the planet's biosphere for use by a rapidly growing number of humans for their pleasure and profit alone—has now resulted in a crisis in the governance of human lives, a crisis of bio-power itself. More importantly, it has brought into view the connections, or rather the entanglements, that exist between our lives and the deep, evolutionary history of microbes.

The pandemic is thus not an event in our global history alone. It is not merely an example of the Great Acceleration of human flourishing. It is also an event that shows, in the form of the unfolding of a drama often tragic for humans, how our increasingly global existence reveals to us the deep-historical (or planetary) aspects of our lives. The novel coronavirus is evolving. What we hear about the Delta variant, or of other variants of the virus, is about its biological evolution. Everything we throw in the path of the virus to disrupt its journey has the potential to become an evolutionary pathway for the virus. The human body itself is now one such pathway. Lynn Margulis and Dorian Sagan reminded their readers some decades ago that

> our species [are] not ... lords but ... partners: we are in mute, incontrovertible partnership with photosynthetic organisms that feed us, the [microbial] gas producers that provide oxygen, and the heterotrophic bacteria and fungi that remove

oxygen and convert our waste. No political will or technological advance can dissolve that partnership.[46]

Researchers of infectious diseases have long been aware of this aspect of the deep and always-present history of humans. Morens, Folkers, and Fauci opened a 2004 article examining the challenge of emerging and reemerging infectious diseases by remembering the warning that Richard Krause, the director of the US National Institute of Allergy and Infectious Diseases from 1975 to 1984, issued in his 1981 book, *The Restless Tide*, that "microbial diversity and evolutionary vigor were still dynamic forces threatening mankind."[47] They ended their article by referring to the role that the evolution of microbes played in the history of infectious diseases: "Underlying disease emergence are evolutionary conflicts between rapidly evolving and adapting infectious agents and their slowly evolving hosts." They added, "These are fought out in the context of accelerating environmental and human behavioral alterations that provide new ecological niches into which evolving microbes can readily fit." This is an ongoing, unending battle in which humans are forced constantly to improve and upgrade their medicines and technology while the microbes evolve and manage, often in situations precipitated by human actions, to switch hosts. Morens and Fauci conclude their own essay by observing,

> The challenge presented by the ongoing conflict between pathogenic microorganisms and man has been well summarized by a noted champion of the war on EIs [emerging infections], [the Nobel Laureate] Joshua Lederberg, "The future of microbes and mankind will probably unfold as episodes of a suspense thriller that could be entitled *Our Wits Versus Their Genes*."[48]

Morens and Fauci returned to this theme in their recent reflections on the current pandemic: "In the ancient ongoing struggle between microbes and man," they write, "genetically adapted microbes have the upper hand in consistently surprising us and often catching us unprepared."[49] Even the technologies we invent to fight microbes generally end up creating new pathways of infection and evolution. Invented in the 1930s, antibiotics gave rise to the feeling in the 1960s that—as Krause put it—

> there seemed little left to do in the battle against infections other than begin a mopping-up operation. It appeared that only a few stubborn serious infections resisted the two-pronged attack of antibiotics and vaccines. No one antici-pated the microbe guerilla actions that were to break out from enclaves in the rear....[50]

And there lies the story of antibiotic-resistant bacteria. "For example," wrote Krause in the early 1980s, "it takes 40 times as much penicillin to treat some infections today as it did to treat those same infections when penicillin was introduced during World War II." He asked wearily, "What of the future if bacteria can elude our best efforts in this fashion?"[51]

Medical strategies for fighting microbes end up as stories of their evolution. "The emergence of novel pathogens," write the virologist Nathan Wolfe and his colleagues, "is now being facili-tated by modern developments exposing more potential human victims and/or making transmission between humans more efficient than before."[52] They mention how methods of blood transfusion acted as avenues for the spread of hepatitis C, the commercial bushmeat trade circulated retroviruses, industrial food production spread bovine spongiform encephalitis (BSE), international travel spread cholera, intravenous drugs spread HIV, vaccine production caused outbreaks of simian virus 40—all these and other similar developments leading to, among other consequences, "susceptible pools of elderly, antibiotic-treated, immunosuppressed patients."[53]

A particular evolutionary advantage that coronaviruses have

over humans, write Morens and Fauci, is the "genetic instability of microorganisms allowing rapid microbial evolution to adapt to ever-changing ecologic niches." This, they say, is especially "true of RNA viruses such as the influenza virus, flaviviruses, entero-viruses, and coronaviruses, which have an inherently deficient or absent polymerase error-correction mechanisms [no proof-reading capacity, in other words, as they reproduce themselves] and are transmitted as quasi-species or swarms of many, often hundreds or thousands of, genetic variants," a fact that makes it difficult for humans to fight them.[54]

This is fundamentally an evolutionary struggle. It reminds us that humans, the species called *Homo sapiens*, for all their mastery of technology, are not outside of the Darwinian his-tory of life and evolution that unfolds on this planet. Infectious diseases in humans are about microbial survival "by [their] co-opting certain of our genetic, cellular, and immune mechanisms to ensure their continuing transmission." In making this point, Morens and Fauci refer to Richard Dawkins: "Evolution occurs on the level of gene competition and we, phenotypic humans, are merely genetic 'survival machines' in the competition between microbes and humans."[55] Human flourishing leads to the degra-dation of the environment. This creates opportunities for coro-naviruses of various strains to switch hosts by moving from their reservoir hosts to various mammalian species, whereby they get preadapted to human cells by working inside other mammalian bodies. Morens and Fauci write, "viruses have deep evolutionary roots in the cellular world."[56] They add, "Evidence suggests that there are many bat coronaviruses pre-adapted to emerge, and possibly to emerge pandemically."[57]

Infectious diseases are about the evolutionary connections that exist between our bodies and other bodily forms of life (one reason why we can develop vaccines by testing them first on other animals). Zoonotic pathogens are responsible for 60 percent of human infections and are "those that *presently and repeatedly* pass between humans and other animals"; the other 40 percent, including smallpox, measles, and polio, "are caused by pathogens descended from forms that must have made the leap to human ancestors sometime in the past." Quammen, from whose book *Spillover* I have cited these words, makes a telling point about the dotted-line relationships that connect human bodies to other mammalian bodies through which these microbes travel: "It might be going too far to say that *all* our diseases are ultimately zoonotic, but zoonoses do stand as evidence of the infernal, aboriginal connectedness between us and other kinds of host."[58]

Krause's rhetoric of a permanent war between humans and microbes seems outdated and wrong. But his other question, "What is the nature of this microbial sea, constantly lapping at the shores of man's dominion?" still resonates.[59] "It may be a matter of perspective [as to] who is in the evolutionary driver's seat," remark Morens and Fauci—microbes or humans. Microbial forms of life have persisted on this planet for 3.8 billion years. *Homo sapiens* have been around for 300,000 years. "This perspective," say Morens and Fauci, "has implications for how we think about and react to emerging infectious disease threats."[60]

Provincializing the Political

Something of an unstated assumption in the constitution of the urban and global modern—to borrow the language of Latour—has broken down when we cannot any longer acknowledge and at the same time contain the microbial world in the domain of the phatic.[61] Our sense of everyday time changes when we become acutely aware of the presence of microbes that can harm us. Microbes, however, are the oldest and the most important inhabitants of the planet, and they play a far more critical role in the maintenance of life on it than humans have ever done or ever will. (If anything, we have created the prospect of another great extinction of life.) "The overwhelming majority of life on Earth is microbial!" writes Paul Falkowski in his book, *Life's Engine: How Microbes Made Earth Inhabitable*. He goes on to say, "In fact, there are far more species of microbes than there are of plants and animals combined."[62] In her introductory book on viruses, Dorothy Crawford writes, "Microbes are by far the most abundant life form on Earth. Globally, there are about 5×10^{30} bacteria, and viruses are at least ten times more common—thus making viruses the most numerous microbes on Earth. . . . The oceans cover 65 percent of the globe's surface and, as there are up to 10 billion viruses per litre of sea water, the whole ocean contains around 4×10^{30}—enough, when laid side by side, to span 10 million light years." In addition, they play a vital role in "maintaining life on earth." The oceans' floating population of plankton is made up of viruses, bacteria, archaea, and eukarya. Crawford also points out that one group of planktons, the phytoplankton (plants), "organisms that use solar energy and carbon dioxide to generate energy by photosynthesis," produce almost half of the world's oxygen, the oxygen without which we struggle to survive when infected with SARS-CoV-2 or its variants.[63]

This gives us a glimpse into the ironical nature of the crisis of the biopolitical that we are living through. Bio-power, as Foucault formulated it, was about securitizing human life. Health, food, and housing are part of it. But a frenzied expansion of bio-power over the last several decades—the great acceleration of human history—has undone that security. The story of antibiotics encapsulates this irony. Indiscriminate overuse of these drugs has allowed antibiotic-resistant bacteria to evolve. As Ed Yong puts it in his book on the human microbiome: "Much of modern medicine is built upon the foundations that antibiotics provide, and those foundations are now crumbling."[64] We may have entered an era of pandemics that we will have to match with newer and newer vaccines. Yet we debate only bio-power and sovereignty—as if the virus could still be contained within the phatic—when we debate the politics of pandemic management internal to and between nations. Did Trump or Modi or Scott Morrison mismanage the pandemic? Is Biden managing it better then Trump? Should those who resist vaccination on grounds of religious considerations be treated leniently? These questions are questions of bio-power. From this perspective, the crisis is a failure of bio-power, and questions of income, racial, gender, sexual, nutritional, digital, and other inequalities come up as legitimate issues. We also discuss questions of sovereignty (distinct analytically, as Foucault said, from bio-power) when issues of global versus national management of pandemics are raised and, by implication, the very question of global governance itself receives some attention.[65]

But a larger question from the history of life stares us in the face through this pandemic. *Homo sapiens* are a minority form of life. We could say of humans what Sartre once said about imperial Europeans: They were "nothing more and nothing less than a minority."[66] The microbes, on the other hand, comprise the majority forms of life. They have also been the architects of life on this planet and are central to its maintenance. Their presence inside our bodies makes us what we individually are. They and humans—and there is no human without a functioning microbiome—constitute together a "whole living being" that Margulis, combining three Greek words (*hólos* for "whole," *bíos* for "life," and *óntos* for "being"), referred to as a holobiont.[67]

To think of individual humans and their microbiome as constituting a whole living being is to think about the limits of the received traditions of modern political thought. For that thought has defined the human as a political subject by bracketing—putting in the container of the phatic—the work of deep history, of the geobiology of the planet including the work that microbes do. Our crisis leaves us exposed to a fact that biologists and infectious disease specialists have known for a long time: We are a minority form of life that has behaved over the last hundred or so years as though the planet was created so that only humans would thrive. If all forms of life were human-like—and we sometimes do use our human imagination to think our way into the experiential-moral worlds of animals and birds (think of the imaginative philosophical work of Vinciane Despret)—then humans would be like Whites in South Africa during the apartheid regime, a racist minority dominating the majority with utterly selfish ruthlessness and imperiling everybody in the end.[68] We would wonder if it were possible for humanity as a whole to look on themselves as a "minor" form of life and work toward minoritarian forms of political thought, of the kind that Arendt or Deleuze on Kafka have educated us in, thoughts that would want to avoid "majoritarian"—ironical, in the case of a minority—dreams of domination.[69] If viruses and bacteria were human or human-like, our knowledge of them would look like "colonial knowledge," knowledge of the other that we acquire with a view to—and in the process of—dominating them. Even Yong's otherwise informed and judicious discussion of the human microbiome ends with an all-too-human, a parochially and provincially human, dream of "controlling" them "for *our* benefit:

> We see how ubiquitous and vital microbes are.…They sculpt our organs, protect us from poisons and food, break down our food,…and bombard our genomes with their genes.… We see how we might start to *control these multitudes for our benefit*, transplanting entire communities from one individual

to another, forging and breaking symbioses *at will,* or even engineering new kinds of microbes.[70]

Yong wrote these words before the pandemic broke. If there is anything the current moment of the Covid-19 pandemic has taught us, it is that such a Promethean understanding of what it means to be human is seriously misplaced. The pandemic speaks not only to the global history of capitalism and its destructive impact on human life but also represents a moment in the history of biological life on this planet when humans are acting as the *amplifiers* of a virus whose host reservoir may have been some bats in China for millions of years. Bats are old creatures; they have been around for about fifty million years, and viruses for much, much longer. In the Darwinian history of life, all forms of life seek to increase their chances of survival. The novel corona-virus has, thanks precisely to the intensification of bio-power of the humans, jumped species. It has now found a very effective agent in humans that allows it to spread world-wide. And that is because humans, very social creatures, now exist in very large numbers in big urban concentrations on a planet that is crowded with them, and most of them are extremely mobile in pursuit of their life-opportunities. Our history in recent decades has been that of the Great Acceleration and the expansion of the global economy in the emancipatory hope that this will pull millions of humans out of poverty. Or at least that has been the moral justification behind the rapid economic growth in certain nations in Asia, Africa, and Latin America. From the point of view of the virus, however, the environmental disturbances this growth has caused, and the fact of human global mobility, have been welcome developments. This is no doubt the pandemic is also an episode in the Darwinian history of life. And the changes it causes will be momentous both in our global history and in the planetary history of biological life.

The pandemic thus speaks of our being embedded in deep history and of our entanglements with both animal and microbial lives. The virus mediates the latter two. There is, however, a tension between our human concerns with biopolitical forms of power—concerns that are amenable to human politics—and our knowledge of our connections with the microbiome, connections that unfortunately cannot create (at least not yet) an extra- or post-human collective political subject that would be both human and nonhuman at the same time. Yet if the argument that both planetary climate change and the pandemic are problems that arise out of unprecedented expansion of the human realm in the period of the Great Acceleration is accepted, then the question of "What is to be done?" by humans to mitigate "the era of pandemics" is one that naturally arises for us.

This is where it may be useful to recall a point that Latour has made in many places and in different versions, one of the most recent being a passage in his lectures entitled *Facing Gaia*.[71] Human pursuit of wealth and prosperity in the period of the Great Acceleration has amounted to an undeclared war—but on what? Latour writes with his tremendous gifts of imagination: "With the Anthropocene, the Humans are now at war not with Nature but with . . . in fact, *with whom?* I have had a lot of trouble settling on a name for them." He finally decided, putting it "in the style of a geohistorical fiction," that "the *Humans* living in the epoch of the *Holocene* are in conflict with the *Earthbound* of the *Anthropocene*."[72] "Humans" refers to humans as they saw themselves in the late Holocene as separate from Nature while "Earthbound" are the entanglements of the human, the nonhuman, and the planetary that the Anthropocene revealed and of which the former "humans" are an inextricable part. The war, however, cannot be won, for while the Earthbound and the earth are powers that will not dominate, they cannot be dominated either.[73]

We, particularly the human subjects who still pursue modern-ization and act as though we were still in the Holocene, need to practice what Latour has in many places called diplomacy.[74] Since humans and the Earthbound cannot meet as negotiating subjects, I suggest that what modernizing and global humans need to practice is one-sided diplomacy—somewhat akin, in my memory, to the Chinese unilateral withdrawal in their war with India in 1962—by imagining and then implementing a process of scaling back the realm of the human-modern.

Raphael Lemkin, a Polish-Jewish lawyer whose family was destroyed in the Holocaust, coined the word "genocide."[75] Humans are on the verge of committing what sociologist and writer Danielle Celermajer describes using the word "omnicide," the killing of everything.[76] You may legitimately ask, which humans? Why not specify those responsible? Oftentimes, it is possible to do so. You can point to politicians, financial institu-tions, businesses, governmental failures, with reason. There are indeed times when it is easy to identify those who kill, destroy, and maim others intentionally. But, as Celermajer points out, responsibility or culpability is not always easy to assign.[77] Five hundred million wild animals died just in the first month of the Australian firestorms of 2020. Nobody actively schemed it. Most people did not even desire it. But it happened because of the changes that follow from what we call "anthropogenic climate change."

Celermajer tells a story to explain the situation: "When I was growing up, my parents used to play a Bob Dylan song called 'Who Killed Davy Moore?'" Modeled on the children's rhyme, "Who Killed Cock Robin?", the song explained the story of Davy Moore, a boxer who died in the ring when he was thirty. If you remember the song, you will know that the coach, the crowd, the manager, and the gambling man all said, "Not I." And then they explained, as Celermajer puts it, that "they were just doing what it is that they do."[78]

We, the privileged humans of today, do what we do to keep the human realm expanding, behaving as though we believed—even if we don't—that the earth was created so that only humans would thrive. We all partake—in unequal degrees—of the changes that the Great Acceleration induced in the human condition. Anthropogenic climate change and the pandemic are connected to that acceleration. It is up to us humans to find ways to scale the human realm back without losing sight of the questions that speak either to issues of intra-human injustice or to those of the inextricable entanglements of the human with the nonhuman, Latour's figure of the Earthbound.

2

The Historicity of Things, including Humans

The Anthropocene is a slice of a geological and nonhuman measure of time. In the periodizing schema used to describe the earth's geological history, a geological epoch, the smallest unit of measurement, can last for tens of millions of years. If geologists agree to formalize the Anthropocene, then the Holocene will be one of the briefest epochs in the geological history of the earth. We may already be in the Anthropocene, but we *Homo sapiens* may cease to exist long before the Anthropocene runs out. The Anthropocene, in that sense, cannot be a periodizing device for human history, or we would have to create two Anthropocenes, one covering the period of human existence and the other for a post-human period in Earth's history. If the sixth Great Extinction really happens in the next 300–600 years, we may even have to upgrade the name to that of an era, the Anthropozoic era. These are not categories for periodizing human history even if we accept—as I do—Earth system scientists' proposition that, thanks to population numbers, technology, geographic spread, and domination over other forms of life, humans today have acquired the dubious glory of being a geological force on this planet.

The time of modern human history—as Koselleck famously said—is constituted by our position somewhere in between our experiences and the ever-expanding horizon of expectations and concerns. This may be why some scholars have translated the Anthropocene into the more humanocentric categories of the Capitalocene, Plantationocene, Econocene, or such like terms to see the present as an extension of modern times. But these moves, to my mind, miss out on what I think of as the real challenge of the Anthropocene: It makes humanist historians confront not only the work of the deep, geobiological history of the planet, but also, even more importantly, the historicity of humanity as a "thing." It is, of course, true that as individuals we are both products of deep history and immersed in it. No human artifacts—from the ordinary pen to the space shuttle— are made without the assumptions that humans have opposable thumbs or that they have binocular vision. We may "own" our bodies (as a condition of freedom), but it took millions of years of evolution to design that body. For humanist historians, such facts usually have a taken-for-granted status and constitute a part of the givenness of the world; they belong to specialist scientific knowledge.

This givenness of the world is now breaking down. Everyday news is making us aware that by using fossil fuels and emitting greenhouse gases, by our relentless cutting down of forests, and by our increasing consumption and urbanization, humans are now warming up the planet, acidifying the seas and raising their levels, making the cities hotter, ushering in an era of pandemics, and possibly precipitating the sixth Great Extinction of life. We now realize that humans, with their technological capabilities, are much larger in their impact on the planet than we once imagined and that the planet—or, the Earth system— in turn is much more finite than it once appeared to humans. Humans, in other words, have themselves become a thing-like entity, a nonhuman planetary force that can change the geobiology of the planet.

Yet, until very recently, and with a few notable exceptions (most of whom have been scholars of the human environment), historians of our time have evinced indifference, even strong resistance, to the Anthropocene hypothesis and its attendant vision of humanity as a planetary force or a "thing-like" entity. A case in point is an essay entitled "Consigning the Twentieth Century to History: Alternative Narratives for the Modern Era" that the respected Harvard University historian Charles S. Maier published in the *American Historical Review* in the year 2000 in a gesture of saying farewell to the previous century.[1] It is an erudite and thoughtful article, trying to see what structural changes, moral questions, and issues of periodization the passing century had thrown up for historians. But surprisingly, published twelve years after the Intergovernmental Panel on Climate Change (IPCC) had been set up by the UN, there is not a word in the essay on global warming as a cultural fact of the twentieth century. Maier ponders some of the moral questions that the century posed to the West, reminding his readers of humanity's "dark historical passage" through two world wars and genocides and of Isiah Berlin's remark that this was "the worst century that ha[d] ever been"; he observes with a refined sense of nuance and irony how "modernity" would have had different meanings for the likes of Adorno and Horkheimer and the leaders of postcolonial nations of Asia and Africa.[2] But missing from the large visual field of this essay is any discussion of what may be now regarded as the biggest problem of the twentieth century, cutting across the East-West divide: anthropogenic climate change and the onset, according to Earth system scholars, of the Anthropocene.

One of the missed opportunities, in retrospect, of Maier's essay must be its discussion of the themes of "delay and acceleration" in modern history. Maier disagrees with Koselleck on a particular point: "Our modern concept of history," Koselleck wrote, "has initially proved itself for the specifically historical determinants of progress and regress, acceleration and decay."[3] Maier demurs: "But acceleration is not a sufficient condition for ascribing *some epochal quality to the century*."[4] But that demurral also points to the greatest irony about Maier's essay, since, barely a couple of decades into the twenty-first century, we have more or less accepted the point that the second half of the previous century was mainly about what climate scientists and historians today call the Great Acceleration. And not only that, but also that with the Great Acceleration came an epoch—vast in terms of human time but brief for geological periods—the epoch of the Anthropocene. There was indeed an *epochal* quality to the second half of the twentieth century, the beginning of a very large epoch in fact, one that might even see our civilization out.

How did a historian of Maier's gifts and erudition miss the news about climate change? Is this an instance of the occupational hazards historians face in making sense of their own times? There were clearly some political and philosophical questions of history that shaped Maier's essay, in particular his concern with the futures of democracies and the twin-threats of populism and authoritarian rule, concerns that very much resonate even twenty years after they were expressed. Those were indeed the humanist concerns of the late twentieth century. Coming at the end of the twentieth century, Maier's essay reminded us of the extent to which the very ideas of modernity, modernization, and democracy—and the associated question of human freedoms— occupied theorists and practitioners of history as they took various postcolonial and anti-Eurocentric turns in the last few decades of the twentieth century. While any naive trust in history's capacity to deliver "freedoms" was seriously depleted by the end of the last century, human, and even historiographical, struggles seemed to make sense only when seen as part of a larger and collective effort—however frustrating at times—to make the world more democratic and modern and to do so in plural and polycentric ways. Was such an intense focus on emancipatory and democratic visions of the future one factor that blinded historians to the way the very idea of "human futures" was being put under a cloud by the planetary problem of climate change and the Anthropocene?

Postcolonial Histories and the Emancipatory Visions of Modernity

The last four decades of the twentieth century were times when the marginalized and the hitherto unrepresented in historical narratives—the working classes, women, Indigenous peoples, and other subaltern groups—were allowed into the academic halls of history.[5] History was democratized by being made more inclusive just as students from disadvantaged backgrounds began to enter an expanded higher education sector throughout the world. The concept of modernity was central to historiographical debates during these years; it underwent a noticeable change of fortune in the final three or four decades of the last century, extending into the beginning of this one. Eurocentrism or Western-centric histories lost their appeal and a sense of discomfort about periodization based on a Eurocentric idea of modernity was by then global.

The sentiment was given a powerful scholarly impetus by Kathleen Davis's sustained and searching examination of categories like "medieval" or "feudal" in her book *Periodization and Sovereignty*.[6] My teacher, the Indian Marxist historian Barun De, made the following remark in 1976 and escaped unscathed: "It is possible," he wrote, "that some future historians... might put the 19th and early 20th centuries [in India] at the end of a medieval period of uncertainty, instead of the beginning of the modern period, which still awaits us in the third world."[7] Such a statement would have seen as hopelessly Eurocentric in the 1990s and 2000s. For, the argument then went, if someone were "modern," then they were so with regards to somebody who was not. That "somebody" could soon come to be seen as "backward" or "pre-modern" or "non-modern," or waiting to be made modern, consigned, as I put it in *Provincializing Europe*, to the "waiting room of history."[8]

Shmuel N. Eisenstadt and Wolfgang Schlucter introduced an issue of *Daedalus* in 1998 devoted to the question of "early modernities" by noting that the idea of modernization as a process that resulted in the economic, political, and cultural "convergence" of the world on a model that was broadly Western, would inspire "substantially less confidence" today [in the '90s] than in the 1960s.[9] The spirit of rebellion against modernism and modernist ideas of modernity was everywhere, in all areas of humanities in Anglo-American universities by the 1990s.[10] The philosopher Kwame Anthony Appiah wrote a sentence in his semi-autobiographical book, *In My Father's House*, that captured the spirit of this revolt: "The modernist characterization of modernity must be challenged."[11] Historians, when they did not abjure the word "modernity," got busy democratizing the use of it, distributing the epithet over a wide period of time (thus the "early modern" period) or between social classes. Others discovered alternative, multiple, and vernacular modernities in an attempt to rid the idea of modernity of all exclusivist and judgmental pretension.[12]

The late Patrick Wolfe spoke for many historians when, sweeping aside all discriminatory divisions between the pre-modern and the modern, he asserted that "colonialism's centrality to the global industrial order" meant that "the expropriated Aboriginal, enslaved African-American, or indentured Asian is as thoroughly modern as the factory worker, bureaucrat, or flaneur of the metropolitan center."[13] The sentiment was noble. But if these disparate figures were all *equally* modern and thoroughly so, then clearly their modernity had little to do with differences in their levels of education, urbanity, or any other forms of cultural capital. In what sense could they then be *equally* modern? Was "modern" then simply a synonym for the "global industrial order," and were those caught up in that order to be treated as "modern" by definition? Historian Sanjay Subrahmanyam argued a related point:

> Modernity is historically a global and *conjunctural* phenomenon.... It is located in a series of historical processes that brought hitherto relatively isolated societies into contact, and we must seek its roots in a set of diverse phenomena: the Mongol dream of world conquest, European voyages of exploration, activities of the Indian textile traders in the diaspora, the "globalization of microbes"...[14]

John F. Richards, the historian of precolonial India, attempted in 2003 yet another defense of the use of the periodizing category "modernity," qualified, in his case, by the addition of the adjective "early." In his essay "Early Modern India and World History," Richards clarified what he meant by early modernity: "[Between 1500–1800] human societies shared in and were affected by several worldwide processes of change unprecedented in their scope and intensity.... I call these centuries the early modern period." As a label, it was better than "Mughal" or "late medieval" for it made India seem less "exceptional, unique, exotic" and less "detached from world history."[15] Richards explained that "early modernity" referred to the "global" developments during the period of 1500–1800: "At least six distinct but complementary large-scale processes define[d] the early modern world." They were: (1) "global sea passages" that led to Europeans' "exploration, mapping, reporting"; (2) "the rise of a truly world economy"; (3)

"the growth of large, stable states and other large-scale complex organizations"; (4) "the doubling of world population"; (5) "the intensified use of land" involving destruction or displacement of Indigenous societies; and (6) "the diffusion of several new technologies—. . . New World crops, gunpowder, and printing—and organizational responses to them throughout the early modern world."[16] C. A. Bayly's magisterial opus, *The Birth of the Modern World, 1780–1914*, extends the logic into the twentieth century by arguing that the rise of the nation-state, "massive expansion" of global links, industrialization and urbanization, etc. signaled changes "so rapid" that they constituted a "step-change in human social organization" and could thus be seen as constituting "the birth of the modern world."[17]

Richards's and Bayly's lists spoke for themselves: expanded communication, growth of states and populations, intensification of the use of land, destruction of Indigenous societies, and diffusion of new technologies. In a word, if one could create a hybrid expression: colonization-modernization. In short, these were years when historians of early modern India gave modernization a long and precolonial—as well as global—past. There was much of value here. It did indeed allow us to see, to use Subrahmanyam's word, how histories were "connected." And it did make a country like India a long-term partner in world history. But in what sense was modernization (i.e., "the global industrial order," to revert back to Wolfe's expression) the same as modernity, and how would it qualify precolonial India for the label "early *modern*"?

A European thinker like Marx would have had no problem in answering this question. For him, modern would *not* have been merely a synonym for industrial production or modernization. He would have had a very specific reason for thinking of capitalism or modernization as marking an advance in human history over what came before it, and that reason would have been part of his philosophic vision of human emancipation. The idea of "emancipation" has many roots, as would be true of its conceptual cognates, "freedom" and "liberty," two similarly inspiring and global ideas in human history. Many of these roots go back at least to the nineteenth century when we hear of the "emancipation of slaves," and later in the century, of "freedom" as conceived in the philosophical traditions of Marxism and liberalism. Jürgen Osterhammel characterizes the nineteenth century as being, among other things, "a century of *emancipation*." But this was a time, he explains, when the word "emancipation," was "derived from Roman law and emphatically European, [and was] far less likely to be applied to the world as a whole."[18] Thus Marx's philosophical-economic category "capital" had inherent in it the idea of juridical equality (through the idea of the legal contracts that wage-labor entailed as well as through the idea of abstract labor) and presaged the figure of the rights-bearing citizen, a step toward his vision of freedom. And in response to today's charge of Eurocentrism, Marx would have agreed that there was indeed a certain precedence given to "bourgeois-European" categories in his philosophy.[19] But Marx's options were not available to a historian who had denounced and abandoned Eurocentrism tout court.

Immanuel Wallerstein once put a useful gloss on the analytical distinction between modernization and modernity with reference to this theme of emancipation for humans:

> The first one is the supposed triumph of humankind over nature, through the promotion of technological innovations. The second one is the triumph of humankind over itself, or at least over oppressive forms of human privilege and authority, through successful resistance to political tyranny, clerical bigotry, and economic servitude.[20]

One could say that the "libertarian" or emancipatory project constituted a certain kind of European self-reflexivity with regard to the first project, i.e., the technological project of modernization.[21] European early modernity matters in part because seventeenth- and eighteenth-century debates in political philosophy remained relevant to all the discussions of modernity that followed them.[22] Hobbes and Spinoza are still pertinent to arguments on democracy in a way that may not find parallels in other instances of early modernity.[23]

Yet while refusing to treat any kind of European *Sonderweg* to modernity as a blueprint or template for understanding the histories of modernization and modernity outside of Europe, the new, late-twentieth century historiography of the non-West still hung on to the idea of the twin emancipation that Wallerstein explained. Hannah Arendt articulated a sense of this legacy in the late twentieth century in one of her posthumously published lectures, "The Freedom to be Free." Arendt postulated, going back to the late eighteenth- and nineteenth-century revolutions, that there is a fundamental relationship between freedom from fear and—in today's terms—freedom from poverty: "To be free for freedom meant first of all to be free not only from fear but also from want."[24] It was this dual-engine of a composite desire for freedom from having to fear "the White man" and for freedom from hunger and poverty that drove the anti-colonial and revolutionary movements in Asia and Africa in the mid-twentieth century.

The Bandung conference of 1955 came out of this teleological view of the future, this new-found and highly unstable vision of a world undergoing more growth and development without any domination of one nation by another.[25] This was also an imagination of a different global-international regime, an imagination fueled by the claims that the colonized made on the European

Enlightenment. Aimé Césaire, for instance, visualized a project of Europeanization of the world sans European domination:

> I maintain that colonialist Europe is dishonest in trying to justify its colonizing activity by the obvious material progress that has been achieved in certain fields under the colonial regime.... [But] Europeanization [of Asia and Africa] was (as is proved by the example of Japan) in no way tied to ... European *occupation*....The proof is that at present it is the indigenous peoples of Africa and Asia who are demanding schools, and colonialist Europe which refuses them; that it is the African who is asking for ports and roads, and colonialist Europe which is niggardly on this score; that it is the colonized man who wants to move forward, and the colonizer who holds things back.[26]

Frantz Fanon perhaps was the most eloquent spokesperson for this imagination: "All the solutions to the major problems of humanity," he wrote, "existed at one time or another in European thought. But Europeans did not act on the mission that was designated them....The Third World is today facing Europe as one colossal mass whose project must be to try and solve the problems... Europe was incapable of finding the answers to."[27] This was a new vision of globality of the anti-colonial, modernizing imagination, an imagination that acknowledged its debt to Europe in a full-throated manner and yet asserted its sovereign, anti-colonial values. Humanocentric, yes, but resolutely anti-imperial. Emancipation, in my sense, began a new non-Western life in the 1950s and '60s.

The growth of the middle classes outside of the so-called West—in China and India and elsewhere in the world—was and is predicated on a stated priority that, though not pursued everywhere with the same vigor, rhetorically continues to hold a very important place in all statements justifying unbridled economic growth. This was the question of elimination of mass poverty. Maria Hsia Chang, an American political scientist who has studied Deng Xiaoping, notes how Deng understood the "fundamental purpose" of Chinese communism to be "the elimination of poverty through the emancipation of the productive forces."[28] One could say the same of Nehru in India whose penchant for dam-building and power-irrigation came out of a desire to feed the "starving millions" who had been subject to frequent bouts of famine during the years of British colonial rule.[29] Mass poverty itself was a product of modernity. Sanitation, public health strategies, medicines including antibiotics, the control of epidemics and pandemics, the green revolution of the late 1960s—measures underwritten by access to cheap energy in the form of fossil fuels—allowed a greater percentage of the poor to survive without any significant improvements in the quality of their lives. Mass poverty emerged as a problem of the first order in these new nations of the 1950s and '60s that economic growth, development, and modernization—each of these terms granting these new nations a sense of where their histories ought to be headed—were meant to tackle. This was a vision of modernization as a path to "emancipation." Listen to Ambedkar, the charismatic leader of the ex-untouchable castes in India, the Dalits:

> Machinery and modern civilization are ... indispensable for emancipating man from leading the life of a brute, and for providing him with leisure and making a life of culture possible.... A society which does not believe in democracy may be indifferent to machinery and the civilization based upon it. But a democratic society cannot [be].... The slogan of a democratic society must be machinery, and more machinery, civilization and more civilization.[30]

The vision became a part, later in the century, of the idea of globalization and remained tinged with an earnestness that was still visible in the time of Deng Xiaoping in China and Manmohan Singh in India (Singh was India's Finance Minister when India liberalized her economy in 1991 and later prime minister). That earnestness transformed into authoritarianism and bad faith with later leadership but—and this is my point—a legacy of "obligation to the masses" remains central to the legitimization that both the Chinese and Indian regimes seek both inside their nations and globally. The rhetoric of the appeal, "We need fossil fuels to move millions out of poverty," has a global force because of the pull the idea of emancipation is still capable of exerting on the conscience of the privileged (compare Bill Gates with Elon Musk as visionaries of the future of technology).

The year 1989 dramatically demonstrated the connected but different visions of freedom and emancipation that marked the West and the East. It was not just about the significance of the fall of the Berlin Wall for Western democracies; it was also about the Tiananmen Square protests of the same year. As Bruno Latour asked in a recent essay, did the philosophers of history who declared capitalist and liberal democracy to be the unquestionable victor of the Cold War fail to see the problem of anthropogenic climate change because they had—on behalf of a self-satisfied West—too much "moral clarity?"[31] But we cannot invoke such moral clarity on the part of the West to explain why intellectuals outside of the West failed to take the climate question seriously. After all, the colonized and the Western critics of European empires had always known how flawed and misleading that "moral clarity" of the West was, since it accompanied and justified every piece of colonial aggression. The late 1980s and the 1990s remain important in the non-Western world and in postcolonial thinking precisely because visions of and growing opportunities for the recognition of the rights of the formerly marginalized opened up new horizons of emancipatory aspirations. Reading

The Wretched of the Earth seemed more meaningful and urgent than reading *Silent Spring*; the story of a high-yielding variety of wheat invented in Mexico seemed headier than the cautionary message of *The Limits to Growth*.[32] True, Latour's *We Have Never Been Modern* came out in 1991, issuing a profound critique of "modernity" and of what he called the "constitution of the modern."[33] But these were also the years when postcolonial criticism, having begun its career in 1978 with the publication of Edward Said's *Orientalism*, was reaching its crescendo with the publication of *Subaltern Studies* in India in 1982, Partha Chatterjee's *Nationalist Thought and the Colonial World*, and Gayatri Spivak's essay "Can the Subaltern Speak?," both originally in 1986, as well as the release of Homi K. Bhabha's *The Location of Culture* in 1994 and Arjun Appadurai's *Modernity at Large: Cultural Dimensions of Globalization* soon after in 1996.[34] My *Provincializing Europe*, published in 2000, was the tail end of this movement.

Postcolonial thought of the 1990s—for all its critique of the nation-state and race-class formations—was just as environmentally blind as anti-colonial modernizing nationalisms. It took modernity and the modernization of the world for granted. Latour's and Latourian critiques of modernity, on the other hand, did not connect with the postcolonial desire for growth, modernization, and the democratization of consumption that the digital revolution enabled. Now it is possible that the critiques mounted by Latour or Philippe Descola, which staged at the center of their narratives a grand anthropological clash between civilizations that were based on a fundamental distinction made between "nature" and "culture" and those—mainly Indigenous—that were not, looked upon Asian, African, and Latin American anti-colonial "moderns" as unoriginal, fading carbon copies of their predecessors in Europe. But postcolonial scholarship—as indeed can be seen in the works of the *Subaltern Studies* group,

Spivak, Bhabha, Appiah, and others—was precisely about resisting any attribution of "derivativeness" or "mimicry" to the emancipatory desires of anti-colonial modernizers. The debate has been much less resolved than forgotten, as we shall see in the following chapter.

At the same time as the air was becoming laden with more greenhouse gases and particulate matter, the size of the consuming classes in the world was growing, people were being pulled out of poverty in large numbers in places like China and, more importantly, descendants of the formerly colonized, the enslaved, the underprivileged, and the marginalized were joining the ranks of these newly propertied classes. Inside the West, the struggle was seen to be primarily against racism and for Black, Indigenous, and other minority peoples' rights, struggles that spoke to an emergent sense of democracy but, sadly, not to any idea of an enveloping planetary crisis of the environment. That awareness would not be visible until the crisis was deeper and, so to speak, upon us. Even in my own thinking in the work I did in association with colleagues in *Subaltern Studies*, the onward march of human history—toward more rights, toward democracies to come—went straight past the world of nature until (speaking for myself) my train of postcolonial thought crashed into the planetarity of the 2000s when, in 2009, I published my essay, "The Climate of History: Four Theses."[35]

The Human-Thing as a Challenge to "Modern" Philosophies of History

Maier's essay of 2000 could be read as signaling an exhaustion of the belief that increased modernization would indeed lead to "more" modernity, the molding of the political sphere by the reflexive capacity for reasoning and judgment. But even in that dark mood of questioning, there glows a certain legacy of the debates on modernity, reason, prejudice, and ideas of what Derrida called "democracy to come."[36] This is the legacy of various emancipatory visions of human history discussed above, but this legacy could be sustained only by considering human history on its own, as it were, in separation and isolation from the history of "nature" or that of "things."

I agree with Latour that the received philosophies of history that emphasized emancipation are in crisis today.[37] Let me clarify, however, that I am speaking of a particular strand of historiography and its inheritances. I am speaking of the historical literature on modernity and modernization in the era of decolonization that followed the end of the Second World War. My use of the expression "philosophy of history" also indexes a very particular heritage of thought that could emerge only after history, the discipline, came to be understood in a modern sense as a general phenomenon, not as history-of-this or history-of-that, as Koselleck pointed out a while ago.[38] I refer to secular philosophies of history that were about divining the nature of the movement of history as a whole. This was different from the act of reading or writing history contemplatively, an exercise that historians sometimes engaged in both during and before the Enlightenment. When Machiavelli read Livy, for instance, he put Livy to the same use as ancients would have made of historical accounts: he read history to learn lessons for individual or collective-political life. Thus, to give an example taken at random from Machiavelli's text, he writes:

> I am reading this History of Titus Livius with a view to profit[ing] by it, I think that all the methods of conduct followed by the Roman people and senate merit attention. And among other things fit to be considered, it should be noted, with how ample an authority they sent forth their consuls, their dicta-

tors, and the other captains of their armies, all of whom we find clothed with the fullest powers. . . . This matter I have dwelt upon because I observe that our modern republics, such as the Venetian and the Florentine, view it in a different light . . . [which has] brought Italy to her present condition.[39]

Even when Gibbon published (in 1776) the first volume of *The Decline and Fall of the Roman Empire* and described Tacitus as "the first of historians who applied the science of philosophy to the study of facts," he did not mean anything like what Kant or Hegel or Marx would later teach us to think of as the philosophy of history.[40] Tacitus was not, in Gibbon's account or otherwise, someone who was out to divine the grand telos of human history—whether in terms of class struggle or the cunning of reason. He was rather someone whose *Germania* allowed Gibbon to think of larger generalizations, such as the impact of climate on national histories or of the distinction between the civilized and the savage. Thus, Gibbon writes: "The Germans, in the age of Tacitus, were unacquainted with the use of letters; and the use of letters is the principal circumstance that distinguishes a civilized people from a herd of savages incapable of knowledge or reflection."[41] This is an example of what Gibbon called "contemplating"—as distinct from simply describing—history, and it helped him in his battles with the antiquarians of his time, battles, as Momigliano noted some time ago, that were important for the emergence of the modern discipline of history.[42]

Gibbon was writing Enlightenment history, and the word "philosophy," when applied to history in this period, acquired a particular meaning, as J. G. A. Pocock showed with great erudition in his recent volumes entitled *Barbarism and Religion*. Pocock recalls that the word "philosophy," as the term was used "towards the end of the seventeenth century, did not always indicate a body of systematic thought about nature and knowledge."[43] It expressed a "civil attitude of mind, an openness to reason, a desire to control the passions of which fanaticism was one; but in this apparently eirenic sense [given the wars of religion], it became the basis of a militant programme and ideology."[44] This developed into an interest in social mores, morals, and the question of civilization in any society. Voltaire, to whom we owe the expression "philosophy of history," wrote a book by that name and intended, writes Pocock, for it to be a preface to his book, *Siècle de Louis XIV*.[45]

To be clear, then, it is not what Voltaire designated "philosophy of history" that is at issue in this discussion. The philosophy of history, the activity of reflecting on the process and movement of human history as a whole—and not discussions of the historian's craft (though the two are connected)—springs from a family of "progressive" ideas about human futures that have dominated the world for more than a hundred years in many different incarnations. We may find diverse roots for this tradition; some of the relevant ones surely go back to Hegel's thoughts.[46] But the tradition really comes into its own in the second half of the nineteenth century in the idea of progress (Marxism being one variant of it) and then appears in different shapes and sizes in the twentieth century under a variety of names such as "industrialization," "modernization" (both socialist and not), "development," and so on. The key intellectual problem here, as Peter Wagner pointed out a while ago, was not so much a matter of history being moved forward by anything like some "iron laws" as it was the problem of human "freedom," the amount of "autonomy" that human reason could find for itself in history.[47] And this necessitated a separation between the history of humans and the history of objects and things. This was a critically important move that we now have begun to question.

Karl Löwith's *Meaning in History* (1949) and R. G. Collingwood's *The Idea of History* (1946) are two mid-twentieth century representative texts on such philosophies. They both recognize that while the expression "philosophy of history" was coined by Voltaire, its meaning changed in the nineteenth century. Löwith in addition sees in post-Enlightenment philosophies of history a certain secularization of Judeo-Christian ideals and arguments against the classical Greeks. This is, of course, a brutally summary way of describing complex texts but what I wish to highlight is the human/thing or the human/nonhuman distinction that postcolonial discussions of modernity and freedom from the 1950s on inherited from identifiable traditions of European thought. The idea of the Anthropocene challenges this key and fundamental distinction of our received philosophical histories of modernity by asserting that humans are also—in a certain mode of existence—a planetary force or a "thing."[48]

Benedetto Croce may be seen, in retrospect, as the patron saint of postcolonial history-writing in the Anglophone world after 1950 since he was the first to propound—at least in the twentieth century—that all history was contemporary history. Simplified and made into a mantra by E. H. Carr, this became a rallying cry for much postcolonial thinking about the past in the late twentieth century (such as in the much-mouthed slogan of the 1970s and '80s, "my experience is my history").[49] The argument that all history was contemporary history could be made only if one assumed that history was an exclusively human enterprise, bound by the very limited sense of time with which humans worked, and that objects, natural or artifactual, could never have history in the same sense as humans did.

The separation of the human sciences from the natural ones took formal, disciplinary shapes toward the end of the nineteenth century, as has already been noted by Jean-Baptiste Fressoz and Fabian Locher.[50] But early justifications for this break are easy to locate in several of Croce's comments. In an essay titled "The

'History of Nature' and History," Croce argued that 'history of nature' was "so only in name," for natural history was mainly an exercise in classification: "The saying that nature has no history is to be understood in the sense that nature as rational being capable of thought has not history, because it is not—or, let us say, it is nothing that is real ... reality is all development and life."[51] Croce acknowledged that "even ... in human history," there existed a "natural history," but one could not convert it back into history. "Do you wish to understand the true history of a Ligurian or Sicilian neolithic man?" he asked. His answer: "If it not be possible, or you do not care to do this, content yourself with describing and classifying and arranging in a series the skulls, the utensils, and the inscriptions belonging to those neolithic peoples."[52] For him, this was like wanting to understand the history of "a blade of grass." "First and foremost," he prescribed, "try to make yourself into a blade of grass, and if you do not succeed, content yourself with analysing the parts and even disposing them in a kind of imaginative history." This is why, he repeated, history was "*contemporary*" and a chronicle (which nature could have) "*past* history."[53]

From this particular essay of Croce's and another published later, we come to know that the distinction between natural and human history was triggered for Croce by the remarks that the German political-economist, Friedrich von Gottl-Ottlilienfeld, made in 1903 at the Seventh Congress of German historians held in Heidelberg. Gottl-Ottlilienfeld's lectures, directed against Karl Lamprecht, were later published under the title, *Die Grenzen der Geschichte* in 1904.[54] Gottl-Ottlilienfeld, in Croce's telling, "energetically denied the community and even the affinity of the historian with the geologist, of whom the [former] has as his object events (*das Geschehen*) and the latter stratifications (*die Schichtung*)" and this difference called for "the emancipation of historical thought from the naturalistic."[55]

Certain debates among British philosophers in the 1930s and '40s accentuated this emphasis on the separation of the history of things and human histories. When the Oxford philosopher Robin Collingwood, a translator and disciple of Croce, wrote his lectures and notes that were posthumously published in 1946 as *The Idea of History*, he argued passionately against views that saw human history as embedded in natural history. Examining his positions gives us some fresh and relevant insights into the intellectual challenges that the Anthropocene hypothesis faces in the humanities and the interpretive social sciences.

We get a sense of the kind of debates Collingwood was involved in by remembering a book that was published a few years after his death by the biologist J. B. S. Haldane, his contemporary and junior by a few years. Haldane published a book of essays in 1951 called *Everything Has a History*. He originally wanted to call the book *The History of England, and Other Essays* and then changed his mind: "A history of England generally turns out to be a few of the better known people in England in each century of the last two thousand years. I think it should mean the history of our Land and should include what we know of the history of its people before any written records were made."[56] Haldane's first essay, "The Beginning," started thus: "In this series of articles I intend to give a brief sketch of the history of England and Wales. By this I mean, not the history of the peoples who have lived in these countries, but of the lands themselves. Scots readers will doubtless complain that I have left Scotland out. There is a good reason for this. The rocks of northern Scotland are much older than most of those in England and Wales. And they have been much more violently disturbed, so that we do not know their history in anything like the detail with which we know those of England."[57]

Haldane's essays had been previously published in popular

magazines such as the *Daily Worker*, the *Modern Quarterly*, the *Rationalist Annual*, *Coal*, and the *British Ally*—probably in the 1940s, so Collingwood, who died in 1943, may or may not have seen them. But even if he had, it can be said without a doubt that he would have roundly objected to Haldane's use of the word "history" to describe changes in the natural world, and to the latter's very extravagant gesture of claiming, against all the principles of Collingwood's philosophy, that everything, yes *everything*, had a history! For in his classic *The Idea of History*, and in the more recent *The Principles of History: And Other Writings in Philosophy of History*—both volumes published posthumously—Collingwood had argued vehemently against just that sacrilegious idea.[58]

I have no evidence that Collingwood actually read these essays of Haldane's, but we can be sure that he had read and commented on Haldane's *The Philosophy of a Biologist*.[59] Collingwood's immediate intellectual adversaries were "philosophers like M. Bergson, Mr. [Samuel] Alexander, and [A. N.] Whitehead," and in particular an essay published by Alexander in 1936 under the title, "The Historicity of Things."[60] In that essay, Alexander argued that "everything is historical, including the world itself as a whole," that human history and the history of things constituted a continuum, that "time enters into the constitution of things," and therefore things possessed what Alexander called "timefulness."[61] Collingwood rejected these observations with inflamed passion: "The great and golden-mouthed philosopher, Samuel Alexander," he wrote, began "his last published essay, 'The Historicity of Things' by saying 'that in his opinion it was time the philosophers went to school with historians.' The lesson he wished them to learn was, in brief, the all-importance of time, 'the timefulness of things.'" The natural world was historical in its construction, full of events, and, Alexander contended, if "history and natural science are in agreement, what ails philosophy that she should stand out of the happy party?"[62]

Collingwood was not prepared to accept this invitation. He was ready to grant that "modern astronomy... gives us a celestial history," and "that modern biology includes among its functions that of a biological history... or... that modern geology is among other things a geological history." He was also aware that "medicine is nowadays interesting itself in the history of disease[s]... or that even physics itself is becoming historical in the mind of a thinker like Whitehead...." But "none of these things," Collingwood insisted, "is history." "Chronology, yes; developments of the age-old idea that nature is essentially process or event, by all means; but history, no."[63] He explained further in *The Idea of History*: "Modern views of nature no doubt 'take time seriously.' But just as history is not the same thing as change, so it is not the same thing as 'timefulness,' whether that means evolution or an existence which takes time." He was sure that if the question of whether history "coincides in essentials with this modern conception of nature" were put to "the ordinary historian," the latter would "answer it in the negative." For according to such a historian, said Collingwood, "all history properly so called is the history of human affairs."[64] This did not mean, however, "that all human actions are subject-matter for history; and indeed historians are agreed that they are not." Collingwood's expanded on this point: "So far as man's conduct is determined by what may be called his animal nature, his impulses and appetites, it is non-historical; the process of those activities is a natural process." A historian, he said, "would not be interested in the fact that men eat and sleep and make love and thus satisfy their natural appetites." But—and this was his point—"he [the historian] is interested in the social customs which they create by their thought as a framework within which these appetites find satisfaction in ways sanctioned by convention and morality."[65]

A Historiographical Conundrum
of the Anthropocene

Because Collingwood mentions activities like sleeping, eating, and making love—activities in which humans engage in as individuals—the word "men" in Collingwood's sentence could only mean an arithmetical sum of individuals (not a collectivity itself as a single entity, as in the expression "dominant species," for example). From what we now know about human microbiomes (human coevolution with microbes) and their relationship to our moods and well-being, this mind-body distinction in the case of human individuals does not seem valid any longer.[66] But the more relevant issue raised for the perspectives on human history discussed here via Croce, Collingwood, and Carr is that of the separation of the history of the human from the history of the nonhuman, the separation that acted as the foundation of what both Löwith and Collingwood called the "philosophy of history."

The untenable nature of this separation has become increasingly self-evident in our time as we acknowledge the capacity that humans have acquired to act as force of planetary proportions. Haldane, Alexander, and Whitehead appear finally to have won the argument in which Collingwood sought to engage them. This does not mean, however, that the concerns that animated philosophies of human history—varieties of emancipatory visions for humans—have ceased to exist. They exist, but only as indices of the predicament that humans have inhabited over the last several decades when, with alarming frequency, we have seen events in global human history doubling up as events in nonhuman histories as well—and the latter on such a planetary scale that the two strands of human and nonhuman histories of humans have seemed inextricably linked.

Latour, François Hartog, and others agree that the Anthropocene is disorienting. In introducing their volume *Critical Zones: The Science and Politics of Landing on Earth*, Latour and Peter Weibel described "the disorientation introduced by the introduction of coal and gas [i.e., geological, deep time] into the rhythm of social and world history":

> Everybody nowadays is aware of the name geologists have given to this disorientation: the *Anthropocene....* The news is so disorienting that every discipline, every interest group offers an alternative term, insisting on this or that other variable, in order to cope with the maelstrom. That's actually the good thing about this new geological label: It has spread everywhere and yet it is impossible to settle quietly "in" the historical period it designates.[67]

Hartog estimates the disorientation in terms of the fragmentation of time itself, "the negative effects" of the Anthropocene "on the world and times of the world." "Our condition," he writes,

> is the result of a number of gaps: the gap between the time of the Anthropocene and the times of the world (which are more and more fragmented), the gap between a digital presentism (which is at the heart of globalization) and the world's other temporalities, and, lastly, the radical gap between presentism and the temporalities of the Anthropocene. This condition is an experiment in dismembering."[68]

At the heart of this sense of dismembering of time and space that the Anthropocene induces in its observers is the fact of entanglement of human time with the deep time of geology, thanks to the growing imbrication of what the art and design theorist Benjamin Bratton calls "carbon and silicon based intelligences."[69] Normally this would not happen. The planet has in the past just seemed too big, almost infinite, for humans to constitute a telluric force. For the latter to happen (i.e., for humanity to act like a physical force of planetary proportions with an impact similar to that of a huge heavenly body), the space-time dimensions of human existence had to become comparable to those of the planet. This could not have happened without an exponential growth in the quality and quantity of technology and its capacity to influence human management of life, both human life and the lives of the animals and birds that humans farm and keep. Humans, technology, and these other forms of life, taken together, form a complex that acts like a hyper-object, a "thing" that does lend itself to directly to human perception. When humans become in their own human history this human-thing, they become much larger in their thing-ly aspects than they were before, while the planet, in comparison, shrinks, appearing much more finite than it once seemed. The historicity of that "thing"—something one may call the "humanity-thing"—cannot be thought of or written from within the assumptions of Collingwoodian and emancipatory thought-worlds. Yet collectively humans today constitute a thing-like entity. In what mode do we exist as a thing or a force? How do we write the history of that mode of being? Should there be a natural history of human modernity? But was not modernity founded on the distinction between the natural and the human? This is the disorientation that any history of our present as part of the Anthropocene must negotiate.

3

Staying with the Present

What do we make of the present, if we do not wish—as Faisal Devji has put it recently—to "lose the present to history?"[1] That is to say, if we do not wish to think of the present as identical to any historical explanations of how it has come to be. There are, of course, many historical explanations of anthropogenic planetary climate change, ranging from factors such as capitalism and empires to industrial civilization. But when we see the present only as an outcome of some past processes, our view of the present gets skewed: It is already oriented to futural questions such as how to reduce greenhouse gas emissions, how to combat capitalism, "becoming" a particular identity, and so on.[2] I do not for a moment deny the importance of these questions and the political tasks they imply. The question I am asking is, How do we orient ourselves to—and stay with—a present in which, as Bruno Latour has pointed out more than once, our knowledge of our planetary environmental crises (including the knowledge of its history and causes) and our inaction in the face of that knowledge go hand in hand? A disorienting present, as we noted in the last chapter. One key aspect of this seeming disorientation that Latour highlighted was the question of what attachment to modernization has historically meant to people in different times and places. He wrote: "*Many people think that we deserve modernization. The kingdom of God, in its secular version—that is, wealth and abundance—is our right*. We are already in the promised land. Why should we leave this promised land? To wander the desert again searching for another? It's unthinkable! Living apocalyptically does not, of course, mean leaving an apocalypse behind, but really to live in the end times . . ."[3]

We do not have to agree with the specific formulation of Latour's. But apocalyptic or not—Latour's use of the word "apocalypse," as François Hartog has suggested, may indeed be Christian and thus bear a particular tradition of imagining end-times—Latour's statement reminds us that one cannot even begin to think about the human predicament called "anthropogenic climate change" without registering the aspirations to modernity and modernization that have driven the politics of development in populous "new" nations like India and China in the last seventy years.[4]

At the heart of the issue of modernization—and thus of modern enjoyments and contemporary forms of human flourishing—is the question of increasing human access to energy. In his Tanner Lectures delivered recently at the University of California, Berkeley, literary critic Michael Warner described electricity—the automatic availability of it at the flick of a switch—as "the unconscious" of modern modes of living.[5] Earth system scientists Mark Williams and Jan Zalasiewicz have recently suggested that "the most striking characteristic of our species, *Homo sapiens*" is that "we have become the energy-grabbing species par excellence, our greed for energy is now being unmatched in all of planetary history. . . ." Perhaps, they say, "Linnaeus should have named us *Homo avidas*. . . . By some estimates, each year we extract two-fifths of all energy available in the bodies of plants visible above ground. . . . To this energy we need to add that supplied—when we eat them—by our domestic animals, like cattle, sheep, and pigs."[6] This human "greed for energy," of course, became more and more visible with the onset of the Industrial Revolution and with the later rise of post-industrial societies. After all, as I have suggested throughout this book and elsewhere, it is only when their technology has reached a certain point of development—and in technology I include the human use of nonhumans, both living and nonliving—that humans collectively (and differentially, both because of and in spite of their internal inequalities) can become a "planetary" force, a very large thing-like entity that can impact the planetary climate system and many other bio-geological processes on scales of time and space that are indeed vast and planetary.

Technology is central to this story. We act as large-scale geological agents or as a planetary force through proxies such as CO_2. A good example of this is the human-technical capacity to produce radioactive waste that can potentially outlive human civilization. It is an example of how technology shows us that planetary time can out-scale the sense of time with which humans operate in their everyday lives. The discovery of radioactivity both extended the impact of human time and contributed to the discovery of deep time by making radiometric dating possible.[7] And if you think, additionally, of the questions how and where humans might dispose of this waste, you will begin to see the spatial and planetary aspects of the reach of modern humans. The two aspects are, of course, connected. Being a geological agent for humans is *simultaneously* a process of becoming and a state of being. We become entwined with the geological—*over human scales of time and space*—whenever we actively enjoy/consume/reproduce the "benefits" or conveniences of fossil-fuel use (in driving or flying or even in the transportation of medicines across the world) or when we passively partake of the historical contributions of coal, oil, and natural gas to our sense of well-being (expanded life-expectancy or increased international transport for goods and people, for instance). But the latter we do in the context of humans having already become a planetary or geophysical force through actants that behave as proxies for humans but that affect things on a planetary scale.

To stay with the sense of a human predicament—inaction in the face of knowledge calling for urgent action—that colors our relationship to *this* energy-intensive civilization, one just has to read together two contemporary commentaries on the benefits that human have derived from processes of modernization and greater access to energy in general: Steven Pinker's *Enlightenment Now* (2018) and John R. McNeill and Peter Engelke's *The Great Acceleration* (2014).[8] At first sight, it would seem that these books could not have been more different from each other. Pinker sees humanity as divided between an Enlightenment party, which worships reason, progress, economic growth, and technical development, and an anti-Enlightenment party, which is opposed to everything that the former party espouses. His "prophets of doom" make up an impressive list: "Nietzsche, Arthur Schopenhauer, Martin Heidegger, Theodore Adorno, Walter Benjamin, Harbert Marcuse, Jean-Paul Sartre, Frantz Fanon, Michel Foucault, Edward Said, Cornell West, and a chorus of eco-pessimists."[9] Pinker's description of Foucault as someone "who argued that the Holocaust was the inevitable culmination of a 'bio-politics' that began with the Enlightenment, when science and rational governance exerted increasing pressure over people's lives" is astounding, to say the least.[10]

The Great Acceleration, on the other hand, is a book filled with the historian's sense of irony and alarm. It does not deny the facts on which Pinker bases his emphatically positive response—at least for the post-Enlightenment world—to Kant's question, "Is the Human Constantly Progressing?"[11] Indeed, if you bracketed the issue of climate change and had no reason to engage with Earth System Science (ESS) that explains how geological and biological pasts of this planet constituted a connected history, you would tend to agree with Pinker that humans have never lived so well as they have, say, since the mid-nineteenth century, and more particularly since the mid-twentieth, thanks to improvements in science, technology, medicine, public health, governmental policies, and so on, following on from the gifts of the eighteenth-century European Enlightenment and in spite of all the growing inequalities between humans.[12]

But anyone with a sense of historical irony would also see that the numbers and graphs that Pinker adduces to show the upward trends in human longevity, numbers, standards of living, and so on could well have been included in what Earth system scientists offer as graphs representing a process of Great Acceleration in human history when both human population and human consumption and standards of living began to increase exponentially in the 1950s and continued to do so through the following decades. The only difference would be that the authors of the Great Acceleration thesis look at this phase of seeming human advances with a rising sense of alarm and anxiety—for this was also the period when humans and their technologies began to interfere profoundly with planetary processes involving the carbon cycle, nitrogen cycle, the hydrological cycle, and other dynamic processes of the planet. McNeill and Engelke's sense of historical irony is palpable, for instance, when they comment, writing on the increase in human population from the mid-twentieth century onward: "After 1945 human demography entered upon the most distinctive period in its two-hundred-thousand-year history. In the span of one human lifetime, 1945 to 2015, global human population had tripled from about 2.3 billion to 7.2 billion. This bizarre interlude, with sustained population growth of more than 1 percent per annum, is of course what almost everyone on Earth now regards as normal. It is anything but normal."[13] Technology and the availability of cheap and plentiful energy have been key to this "success." But this has also meant that humans are now a "planetary force": the "carbon dioxide levels are rising faster than any time known in Earth's history," and "changes to the Earth's nitrogen cycle (through the Haber-Bosch process for the production of artificial fertilizers that keep us alive) may be the greatest in two billion years," and "the scale of trans-continental and trans-oceanic species transfer is a phenomenon without compare in Earth history."[14]

Marshalling the ever-rising figures for human consumption of energy in the last several decades, Williams and Zalasiewicz express a similar sense of alarm: "These extraordinary figures are almost certainly unique in terms of dominance of one species in the 4 billion years of biological evolution. Such an extreme course is not without its dangers. The final line of our own diagnosis might turn out to be, 'The only species in history to knowingly cause its own extinction,' which (among other effects) would leave no one left to read the diagnosis."[15] But let us return to the present that Pinker celebrates.

Pinker is so derisive about the critics of the Enlightenment—and these are mostly venerable Europeans or non-Western critics well known to the West, as if the stakes in this debate concern Westerners more than those outside of Europe or the West—that his prose is unaffected by any sense of self-doubt. Much like many other baiters of the left, he appears to harbor a strong dislike for "intellectuals" because they "hate progress." And it turns out that "intellectuals who call themselves 'progressive'" are even more at fault, for they "*really* hate progress." These intellectuals do not mind using the "fruits of progress" (like the computer) but "the *idea* of progress ... the Enlightenment belief that by understanding the world we can improve the human condition" actually "rankles the chattering class."[16] Pinker's confidence in the progress in modern medicine, for instance, is so overwhelming that it allows him to claim in 2018—six years after the publication of David Quammen's prizewinning book *Spillover*, and with the global experience of swine-flu and other near-pandemics behind him—that humanity has made such "advances" in "economic progress" that they are now "more resilient to natural and human-made threats," so much so that *"disease outbreaks don't become pandemics* [anymore]...."[17] Today that fragment of a sentence stares back at Pinker's reader with the puzzled eyes of an Odradek, the nonhuman character Kafka invented and that Benjamin once described as the form that "things assume in oblivion."[18]

However, it is not my purpose to criticize Pinker with the benefit of hindsight. The point is to mark the strange and remarkable convergence between the two books, Pinker's and the one by McNeill and Engelke, a convergence made possible by the sheer facts of human flourishing they bring together between their respective covers. Even without engaging with ESS—something that McNeill and Engelke set out to do—Pinker cannot but acknowledge that the "ultimate blessing" of increased longevity that modern science, technology, and modes of governance (modernization, in short) have bestowed upon humans has also entailed processes that have now reached a point where some researchers feel constrained to consider the option of engineering the climate of the whole planet in an act of "buying time" for humans as they combat global warming.[19] In other words, whether he says so or not, Pinker's book acknowledges without being aware of the geobiological implications of what he is saying that humans have become a planetary agent or, in my language, a planetary force.

But this is also where Steven Pinker's conditional optimism—we humans would be all right if only "we" did this or that—runs into the problem central to the discussion of this book:

the not-one-ness of humanity. Pinker writes, in opposition to those whom he sees as climate alarmists: "Despite a half-century of panic, humanity is not on an irrevocable path to ecological suicide. The fear of resource shortage is misconceived. So is the misanthropic environmentalism that sees modern humans as vile despoilers of a pristine planet.... Problems are solvable. That does not mean they will solve themselves, but it does mean that *we* can solve them *if* we sustain the benevolent forces of modernity that have allowed us to solve problems so far, including societal prosperity, wisely regulated markets, international governance, and investments in science and technology."[20]

Pinker supports the projects of his Harvard University colleague, physicist David Keith, for "moderate, responsive, and temporary" climate engineering designed "only to give humanity breathing space until it eliminates greenhouse gas emissions and brings the CO_2 in the atmosphere back to preindustrial levels."[21] Pinker is not the only person to produce this conditional, "only-if" statement about human survival. Williams and Zalasiewicz offer another version of it: city-building practices would need to learn not only from Indigenous traditions but also from microbes if cities and their gluttonous appetite for energy were to be a sustainable enterprise. For cities to be sustainable, they write, humans would have to "build *with*, and not against nature." One of the "precedents" they cite comes from the Indian northeastern province of Meghalaya, where the Indigenous Khasi people build enduring wooden bridges "by gradually guiding the roots of the Indian rubber fig tree ... across river ravines, and by trapping roots to tree trunks.... If city developers followed the bridge builders of Meghalaya, they might have at their disposal some of the most dynamic organisms on Earth: microbes, that have had billions of years of practice at designing and contributing to sustainable ecosystems that can include large, robust physical structures" by forming "biofilms that offer protection from dehydration and attacks from antimicrobial materials..."[22] Another example of such a conditional view of the future comes from Indigenous scholar and environmental philosopher Christine Winter, who describes herself as "a Ngāti Kahugnunu, Ngāti Pakeha woman, writing from unceded/stolen Gadigal Country in what is now called Sydney, Australia" in an article that we will consider in more detail. Winter invites us to consider Indigenous ideas about "multispecies justice" and "relationality" in the "what if" form that we have encountered above: "What if multispecies justice [MSJ] could reset social, political, and economic relationships? What sort of futures might MSJ generate?"[23]

All these statements fall like seeds on grounds fertile with human disagreements. Pinker's support for Keith's project has to encounter, for instance, Frédéric Neyrat's considered, humanist, and thoughtful critique of geoengineering that argues for humans acquiring "a capacity for stepping back and regaining some distance [from what they have an impact on]" in a gesture that does not assume a seamless continuity between humans and their "environment."[24] Williams and Zalasiewicz, while recognizing biologically built buildings as technologies that are "nascent," acknowledge that "currently,... despite the best efforts of some individuals and companies, our patterns of consumption remain akin to a parasite on the biosphere." They write, "Diagnosing this condition ... is straightforward; finding remedies is rather less so."[25] Symptomatic of this addiction to energy are solar panels and electric cars, often touted as a grand solution to today's environmental problems. But they are not, say our authors. Batteries for both electric cars and solar panels need rare metals like cobalt and tellurium. These materials are collected from the seabed by using "giant suction devices, or sometimes continuous chains of buckets that rip the delicate ecosystem." If this continues, they add, "the oceans and land will have an impoverished assemblage of organisms for (at least) hundreds of thousands of years," making mass extinction "inevitable."[26]

Williams and Zalasiewicz articulate a dilemma that has its roots deep in certain structures and histories of modernity and modernization. The moral justification for need of access to various forms of energy comes not from affluent societies and classes and their patently self-interested defense of consumerist lifestyles—though there is that problem—but from the need to support the lives of eight billion humans who currently inhabit the planet, with an increasing proportion of them (now half) living in cities. Historically, access to energy led to an increased number of humans, and we now need even more energy—the argument goes—to sustain those numbers. Consider the question of artificial fertilizers. "The extraction of nitrogen from air by the Haber-Bosch process [c. 1909]" is something that Williams and Zalasiewicz liken to "the evolutionary leap forward that led to oxygen-releasing photosynthesis by cyanobacteria deep in Precambrian times. Both processes used ubiquitous materials to increase the available energy in the mass of plant tissues." But there is a crucial difference, they point out: "To make ammonia, it [the Haber-Bosch process] doesn't use the Sun's renewable energy, but the finite resource of fossil fuels ... Some of this artificially fixed nitrogen escapes from soils, to over-feed and hence devastate nearby ecosystems in rivers, lakes, and the sea."[27] Yet, without this environment-destroying technological innovation, the world would have found it difficult to feed its growing number of humans. And the India of my adolescent years would have been stranded with massive food crises, if not famines, in the 1960s if the so-called "green revolution" of the late 1960s had not saved the situation.

Entanglements and Differences:
The Modern, the Late-Modern, and the Non-Modern

Can critique escape the contradictions of our time, our *Jetztzeit*, as it were? There are attempts to stand outside them in the name of a seeming and emergent divide between the postcolonial moment of the last century and the contemporary call in many quarters to decolonize disciplines and the academy. This is how Gianmaria Colpani, Jamila M. H. Mascat, and Katrine Smiet describe this division:

> Here [in decolonial criticism], postcolonial theory is subsumed under Western theory and presumed to share in its flaws. In this vein, decolonial critics often position postcolonial theory as an essentially Eurocentric critique of Eurocentrism. The close association of postcolonialism with poststructuralism and deconstruction is key in this critique.... According to this argument, the close engagement of postcolonial thinkers with authors such as Jacques Derrida, Michel Foucault and Jacques Lacan allegedly turns the postcolonial field into a particular province of a Western and Eurocentric canon. At its strongest, this is even understood as a "betrayal" of its aims.[28]

Déborah Danowski and Eduardo Viveiros de Castro's compelling and thoughtful work gives me an excellent point of entry into this question of the degree to which disciplines and institutions can indeed be "decolonized," i.e., liberated from European or Western thought in the interest of combating planetary climate change. Danowski and de Castro put forward their propositions with exemplary clarity and intellectual vigor in *The Ends of the World*, published in Portuguese in 2014 and in English in 2017.[29] Two issues are central to this book: the problem of multiplicity—the clash between the one and the many—that haunts all politics of anthropogenic climate change, and the contested authority of the (Western) sciences. I appear in the section where they reject the biological concept of "species" (as used in the manner of the recently departed E. O. Wilson) and my deployment of it in the original version of my 2009 essay, "The Climate of History: Four Theses," now revised and reprinted as chapter 1 of *The Climate of History in a Planetary Age*.[30] The actual debate is not relevant here. What I want to highlight are two terms of their critique that are central to the discussion in this book. Here is the passage that contains the two elements I wanted to emphasize—Danowski and de Castro's critiques of modernity/modernization and their political critique of science:

> We must begin by rejecting any sole candidate to the (in)dignity of being the Anthropocene's eponymous. The [E. O.] Wilsonian notion of species is dismissed less on the grounds of its phenomenological evanescence, as in Chakrabarty, than *because it is a tributary of modernity's apolitical, ahistorical, conception of Nature, as well as of the Science's absolute power of arbitrage*. But neither are the revolutionary masses of the classical left, that other recurring incarnation of the modern universal, up to the task; . . . their liberation continues to depend on a generalization and intensification of the modernization front, on the practical (environmental destruction) as well as theoretical (the cult of Nature and Reason) levels.[31]

"Modernity's apolitical, ahistorical conception of Nature" and "Science's absolute power of arbitrage" are identified here as the two objects of criticism. Danowski and de Castro go on to write: "The properly *ethnopolitical* situation of 'human' as intensive and extensive multiplicity of peoples must be acknowledged as being directly implicated in the Anthropocene crisis. If there is no positive human interest, it is because there is a diversity of political alignments among the various world peoples or 'cultures' with several other non-human actants and peoples (constituting what

Latour calls 'collectives') *against* the self-appointed spokespeople of the universal Human."[32]

This line of critique is in continuity with the intellectual program that de Castro had announced in his earlier collection of essays, *Cannibal Metaphysics*.[33] That program was to make anthropology into a *"permanent exercise in the decolonization of thought."*[34] Based on his imaginative reading and analysis of what he called Amerindian "perspectivism" and their "multinaturalism," this decolonizing vision saw both humans and the world as "non-unified," with all prospects of unification lying "in the future, under what we would call a multiple hypothetical mode, and will depend on negotiating capacities once the 'war of the worlds,' as Latour has called it . . . , has been declared."[35]

As de Castro's writings make clear and as he often explains, much of the inspiration for this particular mode of "decolonizing thought" came from the explosive impact that Deleuze and Guattari's work on the ideas of the "savage," the "primitive," the "rhizome," and the nomad had on French thought following the events of May 1968 when France was rocked by a revolutionary upheaval of working-class and student protesters resulting in many weeks of violent civil unrest and economic and political uncertainty and a profound questioning of orthodox Communists. "For my generation," writes de Castro, "the name of Gilles Deleuze immediately evokes the change in thought that marked the period circa 1968, when some key elements of our contemporary cultural apperception were invented. The meaning, consequences and the very reality of this change have given rise to a still-raging controversy."[36] He introduces his own book, *Cannibal Metaphysics*, as one that "puts forward and illustrates a theory of multiplicities—the Deleuzian theme that has carried the greatest repercussions in and for contemporary anthropology," influencing, among others, Latour's critique of modernity in his *We Have Never Been Modern*.[37] As de Castro further explicates, echoing the title of Latour's book, "The concept of multiplicity may have only become thinkable—and therefore thinkable by anthropology—because we are currently entering a nonmerologic, postpopular world where we have never been modern; a world that, more through disinterest than any *Aufhebung*, is leaving in the dust the old infernal distinction between the One and the Multiple that governed so many dualisms, the anthropological pairs and many others as well.... Thinking through multiplicities is thinking against the State."[38] And then again: "Multiplicity is not something like a larger unity, a superior plurality or unity; rather it is a *less than one* obtained by subtraction (hence the importance of the idea of the minor, minority, and minoritization in Deleuze)."[39]

Whether we look at de Castro and Danowski's work or that of Deleuze and Guattari, the Indigenous remains the privileged site and the original instance of this subversive principle of multiplicity, seen as embodying some kind of an absolute Other to the statist ideas of history and modernization that imperial Europe epitomized.[40] "Thinking through multiplicities," writes de Castro, "is thinking against the state."[41] In a significant footnote, de Castro mentions that he wrote this sentence in memory of Pierre Clastres "who was (and remains) one of the rare French anthropologists who knew how to make something out of *Anti-Oedipus*'s ideas, besides being one of the inspirations for the theory of the war machine developed in Plateaus 12 and 13 of *A Thousand Plateaus*."[42] Indeed, one of the pivotal oppositions around which the text of Deleuze and Guattari's *Anti-Oedipus* turned was that between the nomadic and the sedentary. In his preface that described the book as an "introduction to the non-fascist life," Michel Foucault exhorted the reader to "withdraw allegiance from the old categories of the Negative (law, limit, castration, lack, lacuna), which Western thought has so long held sacred as a form of power and an access to reality. Prefer what is positive and multiple, difference over uniformity, flows over unities, mobile arrangements over systems." And in all this, the figure of the nomad that subsumed that of the "savage" or the "primitive" came to occupy a central position; Foucault's injunction to the reader of Deleuze and Guattari was telling: "Believe that what is productive is not sedentary but nomadic."[43]

Deleuze and Guattari opened the famous third chapter of their *Anti-Oedipus*—"Savages, Barbarians, Civilized Men"—by asking, "Where do we find enough innocence" that would allow humans to generate "universal history" after "the universal" had been brought to an end by "the conditions determined by an apparently victorious capitalism?"[44] "Innocence" was not a matter of a dialectical reversal of a binary opposition, not in the way that the idea of a "primitive communism" would be preserved and sublimated into the Marxist ideal of communist society. For Deleuze and Guattari recognized that "universal history" was always "the history of contingencies, and not the history of necessity"; the "primitive system" was self-sustaining, its "death...always comes from without: history is the history of contingencies and encounters."[45] The path back to universal history would similarly include "ruptures and limits," "great accidents...and amazing encounters that...might have never happened."[46] The "primitive" or the "savage," however, supplied a principle critical to the generation of a universal human history, the potential for which capital had destroyed. And hence Deleuze and Guattari's perennial interest in the ethnographic literature on segmentary and acephalic societies. The critical political principle was articulated by placing the nomadic in opposition to the state in their very relationship to the earth. "Only the apparatus of the State will be territorial," write our authors, citing Friedrich Engels, for "it 'subdivides not the people but the territory,' and substitutes a geographic organization for the organization of *gens*."[47] But "where kinship seems to predominate over the earth, it is not difficult to show the importance of local ties." Deleuze and Guattari argued that this is because "the primitive machine subdivides the people, but does so on an indivisible earth where the connective, disjunctive, and conjunctive relations of each section are inscribed along with other relations (thus, for example, the coexistence or complementarity of the section chief and the guardian of the earth)." What states divide, however, is the earth itself: "When the division extends to the earth itself, by virtue of an administration that is landed and residential, this cannot be regarded as a promotion of territoriality; on the contrary, it is rather the effect of the first great movement of deterritorialization on the primitive communes.... Hence the savage, primitive was indeed the only territorial machine in the strict sense of the term ... before there is State."[48]

Ethnographic information about "primitive, segmentary societies" was eventually and famously worked up into the science of nomadology in the twelfth chapter of *A Thousand Plateaus* published in 1980 as the second volume of *Anti-Oedipus*.[49] A part of the chapter—Proposition II—was written in amicable disagreement with but also as "a tribute to the memory" of Pierre Clastres.[50] The starting point once again was the statement that "primitive, segmentary societies" were not only "societies without a State," they were actively organized to keep the state at bay. In disagreement with Clastres, however, Deleuze and Guattari also claimed that such societies did not inhabit "a state of nature" that would enable them to remain untouched by the state. The sedentary and nomadic thus did not constitute a mutually exclusive binary.[51] "The law of the State is not the law of All or Nothing

(State-societies *or* counter-State societies) but that of interior and exterior."[52] There are "huge worldwide machines"—like multinational corporations or religious organizations—that "enjoy a large measure of autonomy in relation to States," and there are also "local mechanisms of bands, margins, minorities, which continue to affirm the rights of segmentary societies in opposition to the organs of State power."[53] Together they constitute the exterior to the state but not a binary outside. And the latter group embodies and illustrates the principles of nomadology.

It is on this terrain of thought—and especially the pace-setting work of Deleuze and Guattari in the wake of May 1968—that the figure of the Indigenous presents itself on the pages of Latour's *We Have Never Been Modern* and in Danowski and de Castro's decolonizing exercise in *The Ends of the World*. Three parties are created in this narrative of a global history of modernity and modernization. I will present them as they are depicted Danowski and de Castro's text that is in deep conversation with Latour's work, predominantly the latter's *We Have Never Been Modern* and *Facing Gaia*.[54] These parties are—in order of the importance ascribed to them and in my terminology—"the original-moderns," "the Indigenous or the non-modern," and "the late-moderns." I am not sure where the enslaved of the north Atlantic would fall in this three-fold distinction, but some of their representatives will turn up in my discussion below. For now, let me stay with this three-fold division.

We know the theoretical and historical lineages of the Indigenous within this threefold schema. They are designated "non-moderns." But who are the original-moderns and why are they "original" in this schema? The original-moderns are North-Western Europeans (including those who moved to the settler colonies), for they are the "Humans of the Holocene" against whom the Earthbound (the living who are opposed to the forces that cause global warming) are up in arms in the geostory that Latour presented in his Gifford Lectures published later as *Facing Gaia*.[55] As Danowski and de Castro gloss Latour's text—these "Humans" are, "it is well understood, none other than the *Moderns*, that race—*originally* North-Western, but increasingly less European and more Chinese, Indian, Brazilian...."[56] The original-moderns are original in two senses. They are the first to become "modern," it is only later that they discover that, "in the East and in the South, other people had learned their lesson too well, taking upon themselves the will and the responsibility for modernization, but in their own, frightful terms."[57] Thus, as modernizers in the East and the South, the Chinese, the Japanese, the Indians, the Brazilian and others are unoriginal in two senses: they are unoriginal in that they come later with Europeans as their predecessors but they are unoriginal also in that they are derivatives, pale copies, as indeed the etymology of the word "original"—from Latin *origo* meaning source or birth—suggests.

Danowski and de Castro are aware of the demographic weight of the unoriginal-moderns compared to numbers of the non-moderns: The 370 million Indigenous people "spread over 70 countries in the world, according to a recent United Nations Permanent Forum of Indigenous Issues (2009) estimate," are "certainly nowhere near the roughly 3.5 billion (read half the human species) crowding our 'technical metropolises,' around a billion of which, it should be noted, live in not particularly 'technical' slums."[58] Yet in spite of their demographic minority—or because of it—the "non-moderns" will carry, in Danowski and de Castro's account, a moral and therefore political weight far out of proportion to their numbers. The reason is simple. The Moderns, original or late, represent a failed project that has now resulted in a catastrophe: "Assured of their privileged access to Nature, Moderns saw themselves as a civilizing force come to convince recalcitrant people to rally to the flag of a common world (a single ontological and cosmopolitical regime) that was also, not by coincidence, the world of the Moderns."[59]

The scientific facts of global warming are not at issue here, for, as de Castro and Danowski say, "we are not discussing if there are such things as global warming and an ongoing environmental collapse; these are among the best-documented... phenomena in the history of sciences.... There is hardly any significant controversy among scientists concerning the anthropic origin of climate catastrophe."[60] The dissemination of this knowledge may even be an "important factor" in bringing people over to the side of the good. But the project of the Moderns cannot unite humanity anymore. "All unification lies in the future," in a post-catastrophic world.[61] The forces for the good "cannot but be an 'irremediably minor' people" (minor in a Deleuzian sense) resembling "less the 'phantom public' of Western democracies than *the people that are missing*, groups that Deleuze and Guattari speak of: Kafka and Melville's minor people, Rimbaud's inferior races, the Indian that the philosopher becomes...—the people, that is, to come; capable of launching a 'resistance to the present' and thus of creating a 'a new earth,' the world to come."[62] It is in "a post-catastrophic time, or, if one wishes, in a *permanently diminished human world*" that "the generally small populations and 'relatively weak' technologies of indigenous peoples and so many other sociopolitical minorities of the Earth could become a crucial advantage and resource."[63]

Now, the question is not whether Indigenous peoples' thoughts and practices could provide both intellectual and practical resources as humans search for a way out of their planetary environmental crises. They, of course, do, and the works of Danowski and de Castro and those of others compellingly show us how.[64] But it is interesting to observe that Danowski and de

Castro's method of effecting a "permanent decolonization" of anthropological thought—much like the Deleuzian tradition from which they take inspiration—does not connect with the emancipatory dreams of not only the late and revolutionary modernizers of Japan, China, India, and Africa, but also someone like Frantz Fanon or, for that matter, B. R. Ambedkar, the greatest modern leader of the Dalits in India, who once publicly asked for Indian society to be completely rebuilt on the principles of liberty, equality, and fraternity![65] Instead, these late-modernizers, by implication considered unoriginal and derivative, are seen as having already been accounted for by the story of the original European-moderns. But would not that amount to ignoring—to continue to speak with Deleuze and Guattari—the ruptures, discontinuities, and contingencies that made modernity what it was in Asia and Africa where lives were impacted no doubt by the domination and racism of European powers but without, in many places such as India, any active elaboration of the near-genocidal logic of European settler-colonial rule?[66] Without that history of Asia (and parts of Africa and Latin America), as we have seen, human history would not have undergone the Great Acceleration or acquired its complexity—what Foucault, always more of a historian than Deleuze and Guattari, called "our immediate and concrete actuality."[67]

I find Danowski and de Castro's demonstration of the entanglement of European and Indigenous thought—both in their own work and in that of Deleuze and Guattari—salutary in the face of certain contemporary discussions that propose too sharp a divide between the European-modern and the Indigenous. They show us the two-way imbrication of European and Indigenous thought in Deleuze and Guattari's texts that look to European ethnography in Latin America, South Asia, and Africa to arrive at the principles of nomadism and multiplicity. Danowski and de Castro, in turn, caught up in the tumult in the world of French thought that May 1968 produced, carry back some of that exciting sense of entanglement into their highly stimulating reflections on Amerindian perspectives on the world—all very helpful in current debates in the humanities about the meanings of climate change.

Yet it is sobering, if not chilling, to hear them say that it is in "a post-catastrophic time, or, if one wishes, in a *permanently diminished human world*" that "the generally small populations and 'relatively weak' technologies of indigenous peoples and so many other sociopolitical minorities of the Earth could become a crucial advantage and resource."[68] This condemns us to a present from which there is no escape except via a catastrophe, no room for the "radical hope" that Jonathan Lear has helpfully distinguished from "mere optimism" which, I concede, is hard to sustain in the face of the sluggishness of nations' responses to the gathering crisis of climate change.[69]

The original-modern, the late-modern, and the non-modern

So, why is the history of the late-modern—the anti-colonial modernizers of Asia, Africa, Latin America, or the Caribbeans—folded into the history of European-Western original-modern as a precondition of Danowski and de Castro's project for achieving a "permanent decolonization of thought" in anthropology? Anna Tsing is right to notice in her imaginative, elegant, and global ethnography of Matsutake mushrooms that "the deployment of the legacy of French structuralism to contrast structural logics has encouraged sharp dichotomies between science and indigenous thought" in certain branches of the human sciences.[70] There is another clue to this intellectual problem that Eric Dean Wilson unwittingly provides in his informative and admirable book on the quandary that air conditioning—and refrigerants in general—pose for the mitigation of global warming.[71]

Wilson acknowledges the historical role of the developed nations (mainly the US) in producing the twentieth-century "hole" in the ozone layer. He also notices, with some alarm, the increasing popularity of air-conditioners in populous nations like China, India, and Indonesia. "And yet I wince, too," he writes, "at the idea that if the demand for middle-class comforts—millions of new air conditioners chief among them—were to match those of the US middle class, parts of those same places—India, China, Indonesia—as well as parts of the rest of the world, could become uninhabitable by the end of the century. There's little evidence to support the belief that a total global transition to renewables could support those unprecedented levels of energy consumption."[72]

However, Wilson is not prepared to grant India, China, or Indonesia as much agency in the matter as he would the developed nations. While it seems understandable that the main target of his criticism should be the profligate use of refrigerants by the United States and other developed nations, the leaders of emerging economic powers appear less responsible for their own decisions in this regard than the Western elite might be. Wilson is sharply critical of the fact that just as "a certain level of comfort and safety became possible for residents of India and China, the United States—along with much of Europe—criticized that desire of the 'developing world' as short-sighted, citing species survival, planetary security, global economic damage, and solidarity, while taking little effective federal action towards its decadent habits, which have been sustained, in large part, by the labor of the citizens of the very countries the United States was bullying to lower their emissions."[73]

Wilson clearly intends to write a global history of the rise of dangerous refrigerants but one that is "just" in its apportionment of blame and responsibility. India and China, because of the size of their populations and with their modernizing drives, will decide the future of global warming. But the past and bigger sinners are the nations of the West, which cannot be absolved of their larger and historical responsibility. "To be sure," grants Wilson, "the desire for an 'American way of life' has spread to the rising middle classes of India and China, some of whom want similar access to climate control and thoughtless energy use. But this doesn't necessarily shift any of the responsibility off US shoulders. *The mold of this particular desire was forged in the West*."[74] And yet again, a few pages on: "We should also remember that US emissions have made—and continue to make—the already hot summers in India, China, and Indonesia even hotter. Heat waves [in]... India [are] making artificial cooling a necessity for mere survival. *What appears to be a growing Indian desire for cooling is at least partly driven by a long history of US emissions*, which have plunged India's already hot climate into certain hell."[75]

This is no doubt a piece of history motivated by a sense of justice, somewhat akin to the "climate justice" position that Sunita Narain and Anil Agarwal took in their 1991 pamphlet, *Global Warming in an Unequal World: A Case of Environmental Colonialism*.[76] But it is also striking that the very idea of justice so imagined requires the historian—Wilson, in this case—to divest Indians and Chinese not only of their historical responsibility for the emission of greenhouse gases (which seems fair) but also of their desires as well. Their "desire" to have access to "climate control and thoughtless energy use" was "forged in the West" and "is at least partly driven by the long history of US emissions." These relatively less significant desires and their histories can therefore be folded into the actions and desires of the United States or the West. This is why the late-moderns do not count for much in this history of planetary environmental crisis. Both their desires and agency can be accounted for by the history of a self-aggrandizing imperial, capitalist, oppressive, exploitative, White, patriarchal, and racist West. But is this not a profoundly imperial gesture that, out of good intentions, ends up denying the entire spiritual history of anti-colonial, emancipatory struggles for autonomy and sovereignty in these places?

Wilson is inspired by the work of Kathryn Yusoff, in particular by her powerful and insurgent text, *A Billion Black Anthropocenes or None*.[77] He asks if "humanity" has ever faced such "a planetary crisis" as today only to recognize the question immediately—in the spirit of James Baldwin—as "hideously loaded."[78] What humanity? Wilson goes on to cite from a passage in Yusoff's book: "If the Anthropocene proclaims a sudden concern with the exposures of environmental harm to white liberal communities, it does so in the wake of histories in which these harms have been knowingly *exported* to black and brown communities under the rubric of civilization, progress, modernization, and capitalism."[79] Note the isomorphism (not identity) of the construction of the arguments: just as the "harms" in question in Yusoff's text were exported under the rubric of civilization and other such like terms, the desires of the Chinese, Indians, and Indonesians were "molded" and "forged" in the West, as Wilson puts it.

Yusoff's polemical title, *A Billion Black Anthropocenes or None*, confronts the reader with the force of an uncompromising either/ or choice—this or none, take it or leave it, no bargaining, no looking for middle grounds of deferral and difference. As Derrida might say, "the title is rather violent, polemical, inquisitorial."[80] But the title betrays the point of this historiography. This is history as warfare, war in the interest of the oppressed who have been violently subjected, expropriated, and/or appropriated to racialized, colonial exploitation of their land and labor. The target is a monist West and its racialized, colonializing academic disciplines like geology or ESS that are supported by a stratum of Black

and Brown bodies whose toil at the coalface produces the knowledge that gives rise to the sciences. It is strategic, then, for this
historiography to ignore the historical differences between three
versions of world-history: that of the settler-colonies that were
founded, indeed, on the dispossession, death, and deprivation of
the Indigenous populations, the histories of those whose bodies
were appropriated through various forms of enslavement, and
those of the colonial middle and laboring classes that the empire
spawned in places like British India. It ignores the engagement
that Eric William, Julius Nyerere, or Gandhi, Tagore, or Ambedkar had with European categories like "capitalism," "socialism,"
"civilization," "representative democracy"—engagements that
gave rise to, as Adom Getachew recently pointed out, alternative
"world-making" practices.[81]

It is in the same strategic interest that this historiography
thinks of the founding violence of any institution as the same as
the violence required for its everyday maintenance: "I want to
suggest that race . . . might be considered as *foundational* to the
production of Global-World-Space and geologic regimes of governance that become manifest in the practices of White Geology (or
the Anthropocene)."[82] In this view, in other words, the founding
violence keeps repeating itself as everyday violence. The contemporary is merely the past repeating itself in its self-same identity.
Hence descriptions such as "(ongoing) settler colonialism," or
as Zoe Todd, an Indigenous feminist scholar from Canada puts
it: "Whereas the European academy tends to discuss the 'postcolonial' in Canada, I assure you that we are still firmly experiencing the colonial."[83] The same strategic interest, I think, requires,
that the category "migrant" be refused. The original settlers were
not "legal" migrants; their dispossession of the Indigenous people was backed by their own state, laws, and the violence they
could mobilize. A legal migrant into a nation founded on (and

maintained by, as this argument goes) settler-colonial violence becomes in turn a "settler." Christine Winter, the Sydney-based Maori philosopher cited above, describes herself as follows:

> I am a Ngāti Kahugnunu, Ngāti Pakeha woman, writing from unceded/stolen Gadigal Country in what is now called Sydney, Australia. This makes me both Indigenous and settler in multiple ways; I must be constantly vigilant to the affects of my positionality. My presence displaces Gadigal presences. My relationships with Gadigal Country are not, and cannot be the same as Gadigal relationships with Country. I am not made of this Country, my ancestors and the spirits of my people do not lie in and walk with this Country. The species of this Country are not my kin: their lives do not entangle with my ancestors' lives, with my whenua.[84]

The ethical sentiments that underwrite Winter's refusal of the category "migrant" are palpable. Besides, her experience of Aotearoa New Zealand as an Indigenous-identified person must also make her sensitive to the Indigenous question in Australia. But where, I wonder, would a non-Indigenous migrant to Australia or Aotearoa New Zealand figure in this historiographical schema? Imagine a Dalit skilled "migrant" from India coming into Sydney. They would surely be beneficiaries of a nation founded on settler-colonial violence and stand to profit from of all the violence that kept the domination of the non-Indigenous in place. But would they become settlers too, even if they were trying to escape the Brahminical oppression of a caste-dominated society back home and elsewhere?[85] Or would they be allowed to identify as "legal migrants," something the original settlers never were? Or consider the question of refugees, a prospect that the climate crisis only makes more real. Think of an illegal immigrant—not a UN-certified formal "refugee" but an asylum seeker or boatperson fleeing oppression, violence, or other forces elsewhere in the world, a boat-person who turns up, let us say, on Australian shores—will they embrace the category "settler?" I raise these questions because they are of critical importance to how to think of the present and its relationship to history. Do we see the present as a continuation of the past structures in various forms? In which case, as Devji said, we end up consigning the present to history. Or do we see history as a process involving ruptures, discontinuities, and contingencies of various kinds, as indeed we saw in our discussion of Deleuze?

These issues are not unconnected to the militant historiography that we only glimpsed in its barest outlines in Danowski and de Castro's "decolonizing" project. It is a historiography that mobilizes global history by assimilating the late-modern into the history of the original-modern. But are there ways of engaging this historiography in a conversation from positions outside it, across historical and philosophical differences? For such a task seems urgent to me in any project of staying—in our thoughts and actions—with the present and with all its messy troubles.

Staying with the Trouble:
Modernization and the Human Predicament

In their book *The Cosmic Oasis* Earth scientists Williams and Zalasiewicz approach the present as a predicament for an increasingly modernizing humanity. They tell a story of a yawning gap between where humans are at in their history and where they need to be if catastrophic changes in the planet's climate system are to be avoided.[86] They do not doubt that the present-day consumerist, capitalist way of life is not viable. Tsing's summary of the human situation seems compelling: "All that [modern] taming and mastering [of a nature assumed to be passive and universal] has made such a mess that it is unclear whether life

on earth can continue.... Interspecies entanglements that once
seemed the stuff of fables are now materials for serious discussion among biologists and ecologists, who show how life requires
the interplay of many kinds of beings. Humans cannot survive by
stomping on others.... Women and men from around the world
have clamored to be included in the status once given to Man."[87]

We now know, thanks to the work of many scholars including
Lynn Margulis, Bruno Latour, Donna Haraway, Anna Tsing, and
others, that the "human" of the humanists is an archaic, if not
an outdated, category. Even individual humans exist entangled
with many other living and non-living entities both inside and
outside of their individual bodies. Yet, for all the creative and
powerful scientific fabulations that Haraway, Latour, and others
have given us, we have not found practical paths toward trans-
forming the epistemologically acceptable figure of the human
as an entangled entity into an effective political agent in human
institutions—in parliaments, in the United Nations, in factories,
businesses, or anywhere else. This is not to deny the attempts
that have been made to make room within our institutions for
arguments and positions pertaining to "rights of nature."[88] Our
institutions, however, remain based on the phenomenological
capacities of the human. Not just institutions. Even the minimal
histories of capitalism or racism or masculinity that Tsing or
Haraway offer in their writings—sometimes just by using these
words—do not work unless we imagine humans in their phenomenological aloneness (labor and capital facing each other as
abstract humanocentric categories), experiencing the world with
all the capabilities and shortcomings of the human perceptual
mechanism.

Nothing, however, underwrites de Castro's revolutionary optimism that a climate disaster resulting in "permanently diminished" human capabilities will give humanity yet another chance

at flourishing by making the right use of the accumulated wisdom of the Indigenous non-moderns. That situation may or may not come to pass.

In addition, this ironical world-history of modernization is not without its own share of contingency and accidents. Imagine how the present could have been different if the human population had stabilized in the 1950s or if the world had sourced its energy requirements from nuclear power. Many of our current problems would have still been there with additional crises resulting from the use of fresh water in nuclear reactors, dangers of nuclear proliferation, not to speak of the problem of the disposal of radioactive waste that would have been much more intense. But the warming would have been less. The point is: There is no politics of the planetary predicament of humans without dealing with issues of climate justice that have to do, profoundly, not just with fossilized carbon but also with the emancipatory aspirations that fuel the desire for more and more energy to underwrite "growth and development" in the new, populous nations. It does not get us closer to the reality on the ground if we place the blame for this desire at the door of the United States. Here is a pen-picture of the aspirations of a young village woman from the Shivalik Hills at the border of Haryana and Himachal Pradesh in north India. Radha—"recently married" and in her twenties—spoke thus in 2013 to a Norwegian anthropologist about possible futures for her children:

> Of course, they [her children] will want city life. Life here on the farm is so hard and there is never any rest. In the city you can have a kitchen where you can make food standing up, you can go to the market and eat food in a restaurant.... And in the city, men and women are *barabbar* [equal]. You do not have to veil all the time. Here, if an uncle is coming, you have to veil up and make chai![89]

As this statement demonstrates with crystal clarity, it is not simply the comforts of the body that draw millions to unsustainable cities in India and elsewhere; it is also the dream, however unfulfilled or unfulfillable, of emancipation—in this case, from not just the drudgery but also the rigors of bodily toil and the inequalities and patriarchy of rural lives. There is nothing morally wrong, as such, with humans wanting to live better or longer *so long as they did not imperil themselves in the process.*

Science and technology are part of this aspiration, which does not mean we should not critique them. Consider the second point of Danowski and de Castro's critique: the arbitrage of science. Firstly, it seems to me that we have to acknowledge the pivotal role that the sciences have played in enriching our understanding of how entangled human lives are with other forms of life and non-living entities. We would not have any knowledge of microbial (and majority) forms of life—bacteria, viruses, protists, etc. that escape direct human perception—without the development of scientific instruments and observations since the seventeenth century. As we saw in chapter 1, our battle with the current pandemic is a constant reminder of that fact. Secondly, science, even Earth system science, is part of the historical irony that attends the ultimately tragic history of modernization. Latour and Christophe Bonneuil have helpfully reminded us that there are many ideas about planetarity that are circulating at any one time.[90] This was as true of the past and as it is of the present. Bonneuil, borrowing Hartog's expression "regimes of historicity," makes the additional useful suggestion that just as there have been different traditions of planetary thinking, there have also been dominant regimes of planetarity, planetary ideas that enjoyed the backing of the powers that be in any society.[91] I would also happily concede the point that ESS, given the role it has played in both positing and explaining the anthropogenic origins of the current episode of planetary warming of the earth's surface, represents a dominant regime of planetarity—witness the big-ticket funding that has made this science possible, the backing it has received from the powerful nations of the world, the United Nations, and various other international organizations. This is also what gives this science a touch of irony. It is a product of the Cold War and is dependent on the technological advances that conflict produced. Paul Crutzen, the pioneer of the Anthropocene idea in our generation, once put a positive spin on the irony without, it seems, being aware of the tragic irony of the Anthropocene that his own statement embodied. He said:

> Our negative impacts help us to understand the world. My research on our atmosphere has really terrified me. But finally I thought: What would we have known about the atmosphere if it has not been polluted? Because pollution gave us the impetus and triggered the funding to study the workings of the environment.[92]

Crutzen's statement does not support any Promethean view of human progress—if anything, it inspires the contrary. Nor does it deny that structures of power and domination that determine what kinds of knowledge the sciences will produce—notice his reference to funding. A poorly funded climate science may indeed look different, whence follow the politics of funding. But the knowledge produced still must go through the acceptable procedures and protocols of such production. Every consensus in the sciences exists only to be challenged by new research, which is why scientific consensuses are often harder won than in disciplines in the humanities that thrive on perpetual disagreements and conflicting doctrines and do not even aim for consensus. For all of Yusoff's critique of "White geology" and Bernard Cohn's pioneering critiques of "colonial knowledge," how would one explain the phenomenon of world-class scientists emerging from the ranks of the colonized if the sciences were all constitutively "racialized," "White," and "Western" (which is not to deny any specific instances of racism that non-Western scientists have suffered)?[93] Or how would one explain the pleasures that many Indian historians have taken in practicing precepts of their discipline that surely originated in the West and came to India as part of colonization and domination?[94] Any theory of "false consciousness" or "mimic men" (to remember Naipaul) would be seriously off the mark.[95]

The processes of global modernization do not spring unchanged from a foundational structuralist separation of nature and society that is then imagined as functioning as some kind of original sin, unaffected by the contingencies of history. At the same time, I agree that the differences between today's "decolonial" positions and those once regarded as "postcolonial"—both already always entangled with European thought—cannot be reconciled in some "higher" synthesis. The task, it seems to me, is—to borrow an expression from Haraway and Indigenous philosophers—to make kin, intellectually and across historical differences.

Making Kin, Intellectually

Let me end this book by returning to a proposition I made in the opening chapters: to accept the political as something that is provincially and parochially human. The political is founded on human phenomenology and thus on disagreement. It assumes humanity to be a pluriverse. There is no avoiding this plurality. And yet the intrusion into the human-everyday of a unitary Earth system makes this plurality itself an urgent political issue. It is urgent because it seems likely that humans will not be able to act on the unitary planetary calendar that the various IPCC scenarios of transition and their calculations of global "carbon budgets" assume. It is possible that humans will keep splitting up into "many" the planet that Earth system scientists imagine as "one." But this will also mean that extreme weather events will multiply, as will climate-related tragedies, the brunt of which will be borne by the disadvantaged and the poor of the world. How do diverse and conflicting groups of humans come together around proposed planetary calendars of action? I hope that the idea of making kin with opposing positions or understanding that one may not entirely be in unison with another will be of some help in orienting ourselves in these disorienting times. While there is not a "we" of humans that can function as one political subject, there is much work left to be done on what Mbembe, following Jean-Luc Nancy, calls "being-in-common" in this crisis.[96]

Scholars like Tsing and Robin Wall Kimmerer show the way by pointing to certain intellectual entanglements without which the modern world would not have happened. True, Tsing sometimes draws our attention to boundaries erected between "[scientific] concepts and stories [scientific fabulations]" that, she thinks, have led to an "unfortunate world."[97] But her work also demonstrates the ongoing dialectic between concepts and stories; and, besides, there are many aspects of the physical sciences where the two cannot be separated, as in the science of plate tectonics, for example. True, one cannot deny the genocidal dispossession of Indigenous peoples by settler-colonial aggression and violence and the roles that disciplines like anthropology, history, or even literature played across different forms of colonial domination. A "billion black Anthropocenes" must be acknowledged even as we are called upon to consider the stratigraphers' idea of the Anthropocene. But there are no abyssal gulfs between knowledges that humans have developed at different times and places, though there are, of course, caesuras and fault lines.

Robin Wall Kimmerer's beautiful book *Braiding Sweetgrass* gives a very instructive account of kin-making across different knowledge-practices.[98] She is a trained botanist. We open the book at where she is discussing a "creamy white" mushroom that "pushed up" one morning "from the pine needle duff, out of darkness to light, still glistening with the fluid of that passage. *Puhpowee*."[99] The mushroom makes her think about her relationship to the world as seen through the science of botany and how it is different when she approaches the same mushroom through her "lost" first tongue. Her botanical language is "fluent," as she puts it. She does not reject science. The science of botany gives her a language of "careful observation, an intimate vocabulary that names each little part." Science, she says, "polishes the gift of seeing." But "something is missing.... Science can be a language of distance which reduces a being to its working parts; it is a language of objects." This language is "precise" but "is based on a profound error in grammar, an omission, a grave loss in translation from the native languages of these shores."[100]

Kimmerer's "first taste of the missing language" was when the word *Puhpowee* was on her tongue. "I stumbled upon it in a book by the Anishinaabe ethnobotanist Keewaydinoquay, in a treatise on the traditional use of fungi by our people [the Potawatomi people, who lived around the Great Lakes]." *Puhpowee*, she explained, translated as "the force which causes mushrooms to push up from the earth overnight." "As a biologist, I was stunned," writes Kimmerer:

> In the three syllables of this new word I could see an entire process of close observation in the damp morning woods, the formulation of a theory for which English has no equivalent. The makers of this word understood a world of being, full of unseen energies that animate everything.[101]

This experience led Kimmerer to reclaim/relearn her own language that—like many of the other hundreds of Indigenous languages—had been "washed from the mouths of Indian children in government boarding schools where speaking your native tongue was forbidden."[102]

I do not claim any universality for Kimmerer's thinking. What I claim is exemplarity. She teaches. She shows by her own example how to speak across real and serious differences. Differences that exist within her—her trained biologist-self and herself as an Indigenous person. This is the politics of "being with" in the face of the gathering emergency of climate change. Kimmerer demonstrates that there could be kin-making between the science of botany and Indigenous knowledge of the country. Intellectual kin-making does not erase difference and produce identities. Like entanglements, it allows us to be plural inside, intellectually, inhabiting the dangerous pleasures of what Homi Bhabha once called "difference 'within.'"[103]

Coda

With that small gesture of homage to historical entanglements of postcolonial, decolonial, and European thought, I bring this book to a close. Climate change is a multidimensional, wicked problem that, given the plurality of human interests intrinsic to human history, is not practically amenable to solutions that seem overwhelmingly total: overthrow capitalism or discard modernity. Even leaving aside the self-aggrandizing history of the imperial West, that kind of a position fails to come to terms with the deep and widespread engagement with modernity and modernization that has taken place outside of the so-called West. Yet while totalizing solutions are not very helpful in practical terms, they often allow us to imagine society and life in radically alternative ways. Marx's utopia of a communist society, Gandhi's critique in his *Hind Swaraj* of industrial/Western civilization, are of a kind with intellectual-critical positions now taken under the sign of "the Indigenous"—without these it is hard to imagine how the world could be completely different.[104] These are positions good to think with and from, as Danowski and de Castro's work makes amply clear. At the same time, worldwide investment in modernity and the desire for modernization cannot be separated from certain emancipatory desires that, once European, are now everywhere. But all these historical-cultural fault lines have to be negotiated in an increasingly climate-stressed world where no practical technological measure—from solar panels and electric cars to climate engineering—seems perfect.

Yet the planet is one, at least for Earth system scientists. They regard the planet's climate system as one. Humans are many, divided in multiple ways and yet connected. Making kin is a way of forging connections around and across differences. How quickly we can do that will decide the degree to which human politics will prove itself adequate to the planetary challenges of our time.

Acknowledgments

Working on climate change has left me indebted, intellectually and otherwise, to so many people that it is impossible either to name them all or to describe in any detail the debts I have run up. Let me simply say that the colleagues and friends I thanked in my 2021 book, *The Climate of History in a Planetary Age*, deserve my thanks again as I continue to think with them. In addition, I must now thank scores of friends and colleagues all over the world who have engaged me in critical discussions, formal and informal, of that book. They again are too many to name individually, but it is their probing questions that have helped me think beyond that book.

This book is dedicated in friendship and gratitude to three friends, one of them now sadly departed. Implicit and explicit conversations and arguments with Bruno Latour perhaps mark every page of this book. It is a matter of everlasting sadness for me that he will never see this book in print. Larger than life, he had a rare capacity to draw people into his sphere of friendship while relishing disagreements and disputations. My work benefited enormously from his warm friendship and intellectual generosity. François Hartog, another scholar of immense learning and curiosity and one of the key architects today of our thoughts on historical time, has generously shared his thoughts and writings over the last five or six years, his friendship enriching my life beyond measure. I also take this opportunity to honor the encouragement I have received now for many years from another important thinker and scholar of our times: Étienne Balibar. I have always learned from his work, and I will never forget his critical, generous, and enthusiastic comments on "The Climate of History: Four Theses" when that essay was still in draft. It was he who first prodded me to think about the difference between the concept of a biological species and Marx's idea of a "species-being." And he has been encouraging of this work ever since.

I thank Ramie Targoff and Sue Ramin for the patience with which they have waited for this book while nudging me to finish it. I remain grateful to them for giving me this occasion to expand my thoughts on a set of problems that has absorbed me now for more than a decade. My friend and research assistant Gerard Siarny has shared my intellectual journey for nearly a decade; I thank him for all his help and suggestions, both academic and technical. Words of thanks also go to the reader for the press for their constructive criticism and to Megan Michel Mendonça for her editorial suggestions.

Finally, it is a pleasure to acknowledge the two people in my life without whose love and support my life and work would have been impossible: my wife Rochona Majumdar and my son Arko Chakrabarty. I also remember with affection two other important people in my life, Rimi and Kaveri, whose respective battles with cancer ended, sadly, while work on this book was in progress.

The essays here draw and build upon some materials I have published in the *American Historical Review*, *Contributions to Indian Sociology*, *Daedalus*, and the *Journal of the Philosophy of History*. Chapter 1 was presented as the inaugural annual lecture of the journal *Contributions to Indian Sociology* and was published in 2021 in volume 55, issue number 3, of that journal. It is being published here with some small changes. Chapter 2 builds on and enlarges what I wrote in a printed conversation with Bruno Latour, "Conflicts of Planetary Proportions—A Conversation," in the *Journal of the Philosophy of History*, volume 14, issue 3, in 2020. Chapter 2 also incorporates a few paragraphs from my essay "The Muddle of Modernity," which was published in the *American Historical Review*, volume 116, issue 3, in 2011. And a part of chapter 3 was published as an essay entitled "Planetary Humanities: Straddling the Decolonial/Postcolonial Divide" in *Daedalus: The Journal of the American Academy of Arts and Sciences*, volume 151, issue 3, in 2022. I gratefully acknowledge the permissions received from the editors and publishers of these journals for reusing some of this material here.

Notes

Notes for Introduction: The Planet and the Political

1. Dipesh Chakrabarty, "Postcolonial Studies and the Challenge of Climate Change," *New Literary History* 43, no. 1 (Winter 2012): 1–18.

2. Dipesh Chakrabarty, "The Climate of History: Four Theses," *Critical Inquiry* 35, no. 2 (Winter 2009): 197–222.

3. Nigel Clark and Bronislaw Szerszynski, *Planetary Social Thought: The Anthropocene Challenge to the Social Sciences* (Cambridge: Polity, 2021), 49 and elsewhere.

4. Ian Baucom, *History 4° Celsius: Search for a Method in the Age of the Anthropocene* (Durham, NC: Duke University Press, 2000). I also had sympathetic responses from many, many colleagues (including Baucom) to whom I remain grateful for their encouragement. For references to E. O. Wilson's works in my essay, see "The Climate of History: Four Theses."

5. See chapter 3, "The Planet: A Humanist Category," in Dipesh Chakrabarty, *The Climate of History in a Planetary Age* (Chicago: University of Chicago Press, 2021), 70, 78–79.

6. Ibid., 3–4.

7. Ibid., 69.

8. Ibid., 85; see also 18.

9. *The Climate of History* may be read as an extended elaboration of these seven propositions.

10. Ibid., 70.

11. Erle C. Ellis, *Anthropocene: A Very Short Introduction* (Oxford: Oxford University Press, 2018); Tim Lenton, *Earth System Science: A Very Short Introduction* (Oxford: Oxford University Press, 2016).

12. Clark and Szerszynski, *Planetary Social Thought*, 29.

13. Clark and Szerszynski, *Planetary Social Thought*, 11, 21. Mark Williams and Jan Zalasiewicz, *The Cosmic Oasis: The Remarkable Story of Earth's Biosphere* (Oxford: Oxford University Press, 2022), 29: "The Earth System Science treats the Earth as a unified system comprising—and being more than the sum of—its physical, chemical, and biological components."

14. See Ellis and Lenton, above.

15. www.igbp.net/globalchange/earthsystemdefinitions.4.d8b4c3c12bf3
be638a80001040.html, accessed on January 30, 2022.

16. Jean-Paul Sartre, Preface to *The Wretched of the Earth* by Frantz Fanon,
trans. Richard Philcox, with a foreword by Homi K. Bhabha (New York:
Grove Press, 2004), lix.

17. Kathryn Yusoff, *A Billion Black Anthropocenes or None* (Minneapolis: University of Minnesota Press, 2018), 106.

18. Ibid., 107. See also 12–13.

19. For an expansion of this argument, see Chakrabarty, *The Climate of
History*. For hyper-objects, see Timothy Morton, *Hyperobjects: Philosophy
and Ecology after the End of the World* (Minneapolis: University of Minnesota, 2013).

20. Dipesh Chakrabarty, *Provincializing Europe: Postcolonial Thought and
Historical Differences* (2000; repr. with a new preface, Princeton: Princeton
University Press, 2008).

21. Baucom, *History 4° Celsius*, especially 44–72.

22. Ibid., 22–23.

23. Achille Mbembe, "Decolonizing Knowledge and the Question of
Archive," cited in Baucom, *History 4° Celsius*, 25. This was a lecture given by
Mbembe in 2015. For an online version, see https://wiser.wits.ac.za/system
/files/Achille%20Mbembe%20-%20Decolonizing%20Knowledge%20
and%20the%20Question%20of%20the%20Archive.pdf, accessed 13 October 2022.

24. Thomas Nail, *Theory of the Earth* (Stanford: Stanford University Press,
2021). The following paragraphs draw on what I wrote for a forthcoming
forum on my book *The Climate of History* published in *Environmental Philosophy*. I remain grateful to Jeremy Bendik-Keymer who led and organized
that forum.

25. Nail, *Theory of the Earth*, 95.

26. Ibid., 109.

27. Ibid., 242.

28. Ibid., 242.

29. Ibid., 260 (emphasis added).

30. Chakrabarty, *The Climate of History*.

31. See the symposium in *History and Theory* 60, no. 3 (September 2021) on

François Hartog's essay *"Chronos, Kairos, Krisis*: The Genesis of Western Time."

32. See this press release by the United Nations: www.un.org/press/en/2021 /sgsm20847.doc.htm accessed on January 30, 2022.

33. Aristotle, *The Nicomachean Ethics* in *The Complete Works of Aristotle: The Revised Oxford Translation*, Book V, ed. Jonathan Barnes, trans. W. D. Ross, revised by J. O. Urmson (Princeton: Princeton University Press, 1984), 1,788.

34. Donna Haraway, *Staying with the Trouble: Making Kin in the Chthulucene* (Durham, NC: Duke University Press, 2016).

35. I am grateful to philosopher Jay Bernstein who pointed this out while commenting publicly on a presentation of mine at the New School University in June 2021. His observation that thinking critically about the categories of history was my method of thinking beyond history also seems pertinent in the present context.

Notes for One: The Pandemic and Our Sense of Time

1. Bruno Latour and Peter Weibel eds., *Critical Zones: The Science and Politics of Landing on Earth* (Cambridge, MA: MIT Press, 2020).

2. Jan Zalasiewicz, "Old and New Patterns of the Anthropocene," in "Strata and Three Stories," ed. Julia Adeney Thomas and Jan Zalasiewicz, special issue, *Perspectives: Transformations in Environment and Society*, no. 3 (2020): 16. See also J. R. McNeill and Peter Engelke, *The Great Acceleration: An Environmental History of the Anthropocene since 1945* (Cambridge, MA: Harvard University Press, 2014).

3. Andrew S. Goudie and Heather A. Viles, *Geomorphology in the Anthropocene* (Cambridge: Cambridge University Press, 2016), 28.

4. Clive Ponting, *A New Green History of the World* (London: Penguin Books, 2007), 412.

5. Will Steffen et al., "The Trajectory of the Anthropocene: The Great Acceleration," *Anthropocene Review* 2, no. 1 (2015): 1–18; Ripple et al., "World Scientists' Warning of a Climate Emergency 2021," *BioScience* 71, no. 9 (September 2021).

6. Hannes Bergthaller, "Thoughts on Asia and the Anthropocene," in *The Anthropocenic Turn: The Interplay Between Disciplinary and Interdisciplinary Responses to a New Age*, ed. Gabriele Dürbeck and Phillip Hüpkes (London: Routledge, 2020), 78.

7. For more on this, see the introduction to my book, *The Climate of History in a Planetary Age* (Chicago: University of Chicago Press, 2021).

8. Bergthaller, "Thoughts on Asia and the Anthropocene," 78.

9. For more on this conference, see https://zzf-potsdam.de/en/veranstal tungen/chronopolitics-time-politics-politics-time-politicized-time. For an earlier use of the expression "chronopolitics," see Hartmut Rosa, *Social Acceleration: A New Theory of Modernity*, trans. Jonathan Trejo-Mathys (New York: Columbia University Press, 2015; first published in German in 2005), 12.

10. I am leaving aside the case of those who voluntarily resist vaccination. They are a complex case with many different ideologies informing their choices and their imaginations of the future. But they also surely want to inhabit a "democratic" present in which their choices are recognized and respected.

11. My thoughts here owe a recognizable intellectual debt to François Hartog's discussion of presentism in his books *Regimes of Historicity: Presentism and the Experiences of Time*, trans. Saskia Brown (New York: Columbia University Press, 2015; first published 2003 in French) and *Chronos: L'Occident aux prises avec le Temps* (Paris: Gallimard, 2020), now available in English as *Chronos: The West Confronts Time*, trans. S. R. Gilbert (New York: Columbia University Press, 2022).

12. Nathan Wolfe, Claire Panosian Dunavan, and Jared Diamond, "Origins of Major Human Infectious Diseases," *Nature* 447 (May 17, 2007): 281.

13. David M. Morens et al., "Pandemic COVID-19 Joins History's Pandemic Legion," *mBio* 11, no. 3 (May–June 2020): 1.

14. Ibid., 3–4.

15. Ibid., 3–4 (emphasis added).

16. David M. Morens and Anthony S. Fauci, "Emerging Pandemic Diseases: How We Got to COVID-19," *Cell* 182 (September 3, 2020): 1077 (emphasis added).

17. David Quammen, *Spillover: Animal Infections and the Next Human Pandemic* (New York: W. W. Norton, 2012), 44.

18. R. E. Kahn et al., "Meeting Review: 6th International Conference on Emerging Zoonoses," supplement, *Zoonoses and Public Health* 59, no. S2 (2012): 6.

19. Morens, "Pandemic COVID-19," 4.

20. Quammen, *Spillover,* 40.

21. Ibid.

22. UN Environment Programme and the International Livestock Research Institute, *Preventing the Next Pandemic: Zoonotic Diseases and How to Break the Chain of Transmission* (Nairobi, Kenya: 2020); Barney Jeffries, *The Loss of Nature and the Rise of Pandemics: Protecting Human and Planetary Health* (Gland, Switzerland: World Wide Fund for Nature, March 2020).

23. UN, *Preventing the Next Pandemic*, 15–17.

24. Jeffries, *The Loss of Nature*, 14.

25. Vincent C. C. Cheng et al., "Severe Acute Respiratory Syndrome Coronavirus as an Agent of Emerging and Reemerging Infection," *Clinical Microbiology Reviews* (October 2007): 683.

26. David M. Morens et al., "Perspective Piece: The Origin of COVID-19 and Why It Matters," *American Journal of Tropical Medicine and Hygiene* 103, no. 3 (2020): 955.

27. Quammen, *Spillover*, 207–8.

28. Ibid., 290.

29. Hartog, *Regimes of Historicity*.

30. Pierre Charbonnier, "'Where Is Your Freedom Now?' How the Moderns Became Ubiquitous," in Latour and Weibel, *Critical Zones*, 77.

31. Charbonnier, "'Where Is Your Freedom Now?'"

32. Roman Jakobson, "Linguistics and Poetics," in *Style in Language*, ed. Thomas E. Sebeok (Cambridge, MA: MIT Press, 1960), 5. That Jakobson may not have read much of Bronislaw Malinowski's work and may have taken the idea of the phatic function of language from the Egyptologist Alan Gardiner is discussed in Rasmus Rebane, "The Context of Jakobson's Phatic Function," in *(Re)considering Roman Jakobson*, ed. Elin Sütiste, Remo Gramigna, Jonathan Griffin, and Silvi Salupere (Tartu: University of Tartu Press, 2021), 227–44. The original Malinowski essay is "The Problem of Meaning in Primitive Languages," in *The Meaning of Meaning*, ed. C. K. Ogden and I. A. Richards (London: K. Paul, Trench, Trubner and Co., 1923), 296–336. Also see n. 37 below.

33. In recent email communication (8 October 2021), sociologist Amita Baviskar made the following illuminating comments to me:

I thought you might like to read this Facebook post that I wrote in early May [2021], when we were in the thick of the second wave here in Delhi: "वाच रया? Staying alive?" In the early 1990s, living in the Narmada valley, I found that when Bhil and Bhilala adivasis met acquaintances at the weekly haat [market] or elsewhere, they greeted each other with the enquiry: वाच रया? It was a shortened form of "पुरयि वाच रया?". Literally translated into Hindi, the phrase means "बच्चे बच रहे हैं? Are the kids alive?" Not "How are you?". But this much more basic concern: "Are your kids staying alive?" Among people who lived with hunger and malnutrition, where health care was hardly there, where every mother I knew had watched her infants die, and everything from diarrhoea to snakebite added to the tiny graves dotting a hillside in each village, "वाच रया?" was the right thing to ask. "Are they staying alive?" Because untimely death sat at one's shoulder, a constant companion to life. Little did I think that, 30 years later, I'd be asking this question in my circle of the urban elite. वाच रया?

I am grateful to Professor Baviskar for permission to cite her email.

34. Dipesh Chakrabarty, *The Climate of History in a Planetary Age* (Chicago: University of Chicago Press, 2021).

35. See Arvind Elangovan, *Norms and Politics: Sir Benegal Narsing Rau in the Making of the Indian Constitution* (Delhi: Oxford University Press, 2019).

36. Email from Arvind Elangovan, May 24, 2021. Thanks to Professor Elangovan for permission to cite his email.

37. See the discussion in Gunter Senft, "Phatic Communion," in *Culture and Language Use*, ed. Gunter Senft, Jan-Ola Östman, and Jef Verschueren (Amsterdam: John Benjamins Publishing Co., 2009), 226–33.

38. Michel Foucault, *Security, Territory, Population: Lectures at the College de France 1977–1978*, ed. Michel Senellart, trans. Graham Burchell, English Series ed. Arnold I. Davidson (New York: Palgrave Macmillan, 2007), 115.

39. Latour and Weibel, *Critical Zones*, 75.

40. Foucault, *Security*, 1. The editors of this volume point out (p. 24, n.1) that Foucault had used the expression "bio-power" in his 1975–76 lectures on the principle that "society must be defended."

41. See the brief discussion in Foucault, *Security*, 77–78.

42. Ibid., 22.

43. Ibid., 36, 67–75, 96.

44. Ibid., 276. But then it is true, as Latour remarks, that "when our idea of the position of the Earth in the cosmos is modified, a revolution in the social order may ensue. Remember Galileo: when astronomers declared that the Earth moves around the Sun, it felt as though the whole fabric of society was under attack," Latour and Weibel, *Critical Zones*, 13.

45. Foucault, *Security*, 276.

46. Lynn Margulis and Dorian Sagan, *Microcosmos: Four Billion Years of Evolution from Our Microbial Ancestors* (1986; repr., Berkeley: University of California Press, 1997), 16.

47. David M. Morens, Gregory K. Folkers, and Anthony S. Fauci, "The Challenge of Emerging and Re-emerging Infectious Diseases," *Nature* 430 (July 8, 2004): 242. The full title of Richard Krause's book is *The Restless Tide: The Persistent Challenge of the Microbial World* (Washington, D.C.: National Foundation for Infectious Diseases, 1981). For biographical details on Krause (1925–2015), see David M. Morens, "Richard M. Krause: The Avuncular Avatar of Microbial Science," *Proceedings of the National Academy of Sciences of the United States of America* 113, no. 7 (February 16, 2016): 1,681–83.

48. Morens, Folkers, and Fauci, "The Challenge," 248.

49. Morens and Fauci, "Emerging Pandemic Diseases," 1,078.

50. Krause, *The Restless Tide*, 11.

51. Ibid., 12.

52. Wolfe, Dunavan, and Diamond, "Origins," 282.

53. Ibid.

54. Morens and Fauci, "Emerging Pandemic Diseases," 1080.

55. Ibid., 1078.

56. Ibid., 1980. This is exemplified by the SARS-like bat β-coronavirus, or sarbecoronavirus, whose receptor binding domains appear to be hyperevolving by sampling a variety of mammalian receptors.

57. Ibid., 1081.

58. Quammen, *Spillover,* 137 (emphasis added).

59. Krause, *The Restless Tide,* 12.

60. Morens and Fauci, "Emerging Pandemic Diseases," 1078.

61. Latour famously speaks of "the constitution of the modern" in *We Have Never Been Modern*, trans. Catherine Porter (Cambridge, MA: Harvard University Press, 1993. First published in French, 1991).

62. Paul G. Falkowski, *Life's Engine: How Microbes Made Earth Inhabitable* (Princeton: Princeton University Press, 2015), 39.

63. Dorothy H. Crawford, *Viruses: A Very Short Introduction* (2011; repr., Oxford: Oxford University Press, 2018), 17–18.

64. Ed Yong, *I Contain Multitudes: The Microbes Within Us and a Grander View of Life* (New York: Harper Collins, 2016), 128.

65. On these issues of sovereignty, see the discussion in Nathan Wolfe, *The Viral Storm: Dawn of a New Pandemic Age* (New York: St. Martin's Griffin, 2011), 212–15 and chap. 12. Wolfe writes on the assumption that while more viral storms may indeed be coming, the constitution and assemblage of the powerful institutions of the world will remain the same.

66. Jean-Paul Sartre, Preface to Frantz Fanon, *The Wretched of the Earth,* trans. Richard Philcox (New York: Grove Press, Grove Press, 2004; first published in French 1961), lix.

67. Lena Reitschuster, "Beyond Individuals: Lynn Margulis and Her Holobiontic Wolds," in Latour and Weibel, *Critical Zones*, 353. Yong, *Multitudes*, 157. Reitschuster cites Lynn Margulis, "Symbiogenesis and Symbionticism" in *Symbiosis as a Source of Evolutionary Innovation: Speculation and Morphogenesis*, ed. Lynn Margulis and René Fester (Cambridge, MA: MIT Press, 1991), 1–14.

68. On Despret's work, see Vinciane Despret, *What Would Animals Say If We Asked the Right Questions?*, trans, Brett Buchanan (Minneapolis: University of Minnesota Press, 2016).

69. See Hannah Arendt, "The Jew as Pariah: A Hidden Tradition," in *Reflections on Literature and Culture*, ed. and intro. Susannah Young-ah Gottlieb (Stanford: Stanford University Press, 2007), 69–90. Gilles Deleuze and Félix Guattari, *Kafka: Towards a Minor Literature*, trans. Dana Polen (Minneapolis: University of Minnesota Press, 1986. First published in French, 1975). See also my remarks in "An Interview with Dipesh Chakrabarty" in *Unacknowledged Kinships: Postcolonial Studies and the Historiography of Zionism*, ed. Stefan Vogt, Derek Penslar, and Arieh Saposnik (Waltham, MA: Brandeis University Press, forthcoming).

70. Ed Yong, *I Contain Multitudes*, 264 (emphasis added).

71. Bruno Latour, *Facing Gaia: Eight Lectures on the New Climatic Regime,* trans. Catherine Porter (Cambridge: Polity, 2017. First published in French, 2015).

72. Latour, *Facing Gaia*, 247, 248.

73. Ibid., 281.

74. See the discussion, for instance, in Bruno Latour, *War of the Worlds: What About Peace?* (Chicago: Prickly Paradigm Press, 2002), 36–38.

75. See A. Dirk Moses, "Raphael Lemkin, Culture, and the Concept of Genocide," in *The Oxford Handbook of Genocide Studies*, ed. Donald Bloxham and A. Dirk Moses (New York: Oxford University Press, 2012), 19–41.

76. Danielle Celermajer, "Omnicide: Who is responsible for the greatest of all crimes?" *Religion and Ethics* (blog), Australian Broadcasting Corporation, January 3, 2020, https://www.abc.net.au/religion/danielle-celermajer -omnicide-gravest-of-all-crimes/11838534. The word "omnicide" was coined by the American philosopher John Sommerville (1905–1994) who set up the International Philosophers for the Prevention of Nuclear Omnicide in 1983, https://users.drew.edu/~jlenz/brs-obit-somerville.html, accessed on September 12, 2022.

77. Celermajer, "Omnicide."

78. Ibid.

Notes for Two: The Historicity of Things, including Humans

1. Charles S. Maier, "Consigning the Twentieth Century to History: Alternative Narratives for the Modern Era," *American Historical Review* 105, no. 3 (June 2000), 807–31.

2. Maier, "Consigning the Twentieth Century," 812.

3. Reinhart Koselleck, "History, Histories, and Formal Structures of Time," in his *Futures Past: On the Semantics of Historical Time*, trans. Keith Tribe (Cambridge, MA: MIT Press, 1985), 102.

4. Maier, "Consigning the Twentieth Century," 811 (emphasis added).

5. The first few paragraphs in this section draw upon my essay, "The Muddle of Modernity," *American Historical Review* 116, no. 3 (June 2011): 663–75.

6. Kathleen Davis, *Periodization and Sovereignty: How Ideas of Feudalism and Secularization Govern the Politics of Time* (Philadelphia: University of Pennsylvania Press, 2008). The project has been carried forward in Kathleen Davis and Nadia Altschul, eds., *Medievalisms in the Postcolonial World: The Idea of "The Middle Age" Outside Europe* (Baltimore: Johns Hopkins University Press, 2009). I personally contributed to the debate in a modest way with my essay "Historicism and Its Supplements: Notes on A Predicament Shared by Medieval and Postcolonial Studies" in that volume (109–19).

7. Barun De, "The Colonial Context of the Bengal Renaissance," in *Indian Society and the Beginnings of Modernisation*, ed. C. H. Philips and Mary Doreen Wainwright (London: School of Oriental and African Studies, 1976), 124–25. See also the discussion in chapter 2 ("Subaltern Histories and Post-Enlightenment Rationalism") of my *Habitations of Modernity: Essays in the Wake of Subaltern Studies* (Chicago: University of Chicago Press, 2002).

8. Dipesh Chakrabarty, *Provincializing Europe: Postcolonial Thought and His-*

torical Difference (2000; repr., Princeton: Princeton University Press, 2007), 8. These moves are reminiscent of how European colonizers justified their domination of others by denying the colonized "co-evality," as Jonathan Fabian famously pointed out in his well-known book, *Time and the Other: How Anthropology Makes Its Object* (New York: Columbia University Press, 1983).

9. Shmuel N. Eisenstadt and Wolfgang Schlucter, "Introduction: Paths to Early Modernities—A Comparative View," *Daedalus* 127, no. 3 (Summer 1998), 2. Sanjay Subrahmanyam's essay "Hearing Voices: Vignettes of Early Modernity in South Asia, 1400–1750" in the same issue points out (99) that Eisenstadt himself propounded the "convergence" view in his *Modernization, Protest and Change* (1966).

10. My thoughts here are mainly concerned with non-Western, and in particular Asian, histories. This academic revisionism, of course, does not mean that ideas about modernization have lost prestige in the world of high politics outside the university. As many will remember, the Chinese experiment with capitalism of the last few decades was begun under the banner of "Four Modernizations." The idea is alive and well in much social science writing in Asia (and I assume elsewhere as well). See, for example, Yoshiie Yoda, *The Foundations of Japan's Modernization: A Comparison with China's Path Towards Modernization*, trans. Kurt W. Radtke (Leiden: Brill Academic Publishers, 1995).

11. Kwame Anthony Appiah, *In My Father's House: Africa in the Philosophy of Culture* (New York: Oxford University Press, 1992), 144–45.

12. See, for example, Dilip Parameshwar Gaonkar, "On Alternative Modernities," *Public Culture* 11, no. 1 (1999): 1–18; Charles Taylor, *Modern Social Imaginaries* (Durham, NC: Duke University Press, 2004); Sudipta Kaviraj, "An Outline of a Revisionist Theory of Modernity," *European Journal of Sociology* 46, no. 3 (2005): 497–526.

13. Patrick Wolfe, "Structure and Event: Settler Colonialism, Time, and the Question of Genocide," in *Empire, Colony, Genocide: Conquest, Occupation, and Subaltern Resistance in World History*, ed. A. Dirk Moses (New York: Berghahn Books, 2010), 110.

14. Subrahmanyam, "Hearing Voices," 99–100 (emphasis in original).

15. John F Richards, "Early Modern India and World History," *Journal of World History* 8, no. 2 (1997): 197.

16. Richards, "Early Modern India," 198–206.

17. C. A. Bayly, *The Birth of the Modern World, 1780–1914* (Oxford: Blackwell, 2004), 11.

18. Jürgen Osterhammel, *The Transformation of the World: A Global History of the Nineteenth Century*, trans. Patrick Camiller (Princeton: Princeton University Press, 2014. First published in German, 2009), 916–17.

19. See the discussion in my Preface and Conclusion to *Rethinking Working-Class History: Bengal 1890–1940* (1989; repr., Princeton: Princeton University Press, 2000). Also "Subaltern Studies, Post-Colonial Marxism, and 'Finding Your Place to Begin from': Interview with Dipesh Chakrabarty," in *Contemporary Political Theory: Dialogues with Political Theorists*, ed. Maria Dimova-Cookson, Gary Browning, Raia Prokhovnik (London: Palgrave Macmillan, 2012).

20. Immanuel Wallerstein cited in Alexander Woodside, *Lost Modernities: China, Vietnam, Korea and the Hazards of World History* (Cambridge, MA: Harvard University Press, 2006), 18.

21. Films and film studies have for a long time looked upon the question of "modern life" as precisely a question of the historical actor's time-space experience. See Ben Singer, "Making Sense of the Modernity Thesis," in *Melodrama and Modernity: Early Sensational Cinema and Its Context* (New York: Columbia University Press, 2001).

22. Daniel Carey and Lynn Festa, eds., *Postcolonial Enlightenment: Eighteenth Century Colonialism and Postcolonial Theory* (New York: Oxford University Press, 2009).

23. The literature on this point is vast, but readers may know the degree to which Michael Hardt and Tony Negri's concept of the multitude in their celebrated book draws on Spinozist philosophy: *Empire* (Cambridge, MA: Harvard University Press, 2000). See also Etienne Balibar, *Spinoza and Politics*, trans. Peter Snowden (London: Verso, 1998) and Warren Montag, *Bodies, Masses, Power: Spinoza and His Contemporaries* (London: Verso, 1999). Besides, there is nothing in South Asian history as yet that traces the beginnings of modern disciplinary practices and political thought back to the early modern period, nothing that is comparable to, say, Gerhard Oestreich's *Neostoicism and the Early Modern State* (1982; repr., Cambridge: Cambridge University Press, 2008); Philip S. Gorski's *The Disciplinary Revolution: Calvinism and the Rise of the State in Early Modern Europe* (Chicago: University of Chicago Press, 2003); or John Witte Jr.'s *The Reformation of Rights: Law, Religion, and Human Rights in Early Modern Calvinism* (Cambridge: Cambridge University Press, 2007).

24. Hannah Arendt, "The Freedom to Be Free," *New England Review* 38, no. 2 (2017): 63. I owe this reference to Nazmul Sultan.

25. See my "Legacies of Bandung: Decolonization and the Politics of Culture," *Economic and Political Weekly* 40, no. 46 (2005): 4,812–18.

26. Aimé Césaire, *Discourse on Colonialism*, trans. Joan Pinkham (New York: Monthly Review, 1972. First published in French, 1955), 24–25 (emphasis original).

27. Frantz Fanon, *The Wretched of the Earth*, trans. Richard Philcox (New York: Grove Press, 2004. First published in French, 1963), 237–38.

28. Maria Hsia Chang, "The Thought of Deng Xiaoping," *Communist and Post-Communist Studies* 29, no. 2 (1995): 380.

29. See the discussion in chapter 4 of my book *The Climate of History in a Planetary Age* (Chicago: University of Chicago Press, 2021).

30. B. R. Ambedkar cited in Mukul Sharma, *Caste and Nature: Dalits and Indian Environmental Politics* (New Delhi: Oxford University Press, 2017), 141.

31. See Bruno Latour and Dipesh Chakrabarty, "Conflicts of a Planetary Proportion: A Conversation," *Journal of Philosophy of History* 14, no. 3 (2020): 419–54.

32. Rachel Carson, *Silent Spring* (Boston: Houghton Mifflin Co., 1962); Donella H. Meadows et al., *The Limits to Growth* (New York: Universe Books, 1972).

33. Bruno Latour, *We Have Never Been Modern*, trans. Catherine Porter (Cambridge, MA: Harvard University Press, 1993. First published in French, 1991), 13.

34. Edward W. Said, *Orientalism* (New York: Pantheon Books, 1978); Ranajit Guha, ed., *Subaltern Studies: Writings on South Asian History and Society* (New Delhi: Oxford University Press, 1982–89); Partha Chatterjee, *Nationalist Thought and the Colonial World: A Derivative Discourse?* (London: Zed Books for the United Nations University, 1986); Gayatri Chakravorty Spivak, "Can the Subaltern Speak?," in *Marxism and the Interpretation of Culture*, ed. Cary Nelson and Lawrence Grossberg, (Urbana: University of Illinois Press, 1988), 271–315; Homi K. Bhabha, *The Location of Culture* (London: Routledge, 1994); Arjun Appadurai, *Modernity at Large: Cultural Dimensions of Globalization* (Minneapolis: University of Minnesota Press, 1996).

35. Dipesh Chakrabarty, "The Climate of History: Four Theses," *Critical Inquiry* 35, no. 2 (Winter 2009): 197–222.

36. See Jacques Derrida, *Rogues: Two Essays on Reason*, trans. Pascal-Anne Brault and Michael Naas (Stanford: Stanford University Press, 2005).

37. For an earlier, perceptive statement about this crisis, see Zoltán Boldizsár Simon, "(The Impossibility of) Acting Upon a Story We Can Believe," *Rethinking History* 22, no. 1 (2018): 105–25. For Latour's statement on received philosophies of history, see Bruno Latour and Dipesh Chakrabarty, "Conflicts of Planetary Proportions," *Journal of Philosophy of History* 14, no. 3 (2020).

38. See Reinhart Koselleck, *Futures Past: On the Semantics of Historical Time*, trans. Keith Tribe (1979; repr., Cambridge, MA: MIT Press, 1985), 28.

39. Niccolò Machiavelli, *Discourses on Livy*, trans. Ninan Hill Thomson (1883; repr., New York: Dover, 2007), 245–46.

40. Edward Gibbon, *The Decline and Fall of the Roman Empire* (New York: Modern Library, n.d.), 1:186.

41. Gibbon, *Decline and Fall*, 1:191. See also 1:198.

42. See Gibbon, *Decline and Fall*, 1:194–96 and 1:202, n.71: "Tacit. Germ. c 3 . . . there is little probability that the Greeks and the Germans were the same people. Much learned trifling might be spared if our antiquarians would condescend to reflect that similar manners will naturally be produced by similar situations." Also see Arnaldo Momigliano, "Ancient History and the Antiquarian," *Journal of the Warburg and Courtauld Institute* 13, nos. 3–4 (1950): 285–315.

43. J. G. A. Pocock, *Barbarism and Religion*, vol. 2, *Narratives of Civil Government* (Cambridge: Cambridge University Press, 2001), 18.

44. Ibid.

45. Ibid., 2:107. See also the discussion on Voltaire in Karl Löwith, *Meaning in History* (Chicago: University of Chicago Press, 1949), 104–14; R. G. Collingwood, *The Idea of History*, ed. Jan van der Dussen, rev. ed. (1946; repr., Oxford: Oxford University Press, 2005, 1994), 76–78, 352.

46. See the discussion in Henning Trüper, Dipesh Chakrabarty, and Sanjay Subrahmanyam, "Teleology and History: Nineteenth-Century Fortunes of an Enlightenment Project," in *Historical Teleologies in the Modern World*, ed. Henning Trüper, Dipesh Chakrabarty, and Sanjay Subrahmanyam (London: Bloomsbury, 2015), 3–24.

47. Peter Wagner, "Autonomy in History: Teleology in Nineteenth-Century European Social and Political Thought," in Trüper, Chakrabarty, and Subrahmanyam, *Historical Teleologies*, 324–38.

48. For a selection of texts of the twentieth century gathered together under the category "philosophy of history," see Hans Meyerhoff, ed., *The Philosophy of History of Our Time: An Anthology* (New York: Doubleday, 1959). Also, Karl Löwith, *From Hegel to Nietzsche: The Revolution in Nineteenth-Century Thought,* trans. David E. Green (1964; repr., New York: Columbia University Press, 1991). On humans having a thing-like impact on the planet and its climate, see David Archer, *The Long Thaw: How Humans Are Changing the Next 100,000 Years of Earth's Climate* (Princeton: Princeton University Press, 2009).

49. Benedetto Croce, "History and Chronicle," in *History: Its Theory and Practice,* trans. Douglas Ainslie (New York: Russell and Russell, 1960. First published in German, 1915), 11–26. See the discussion in E. H. Carr, *What Is History?* (1961; repr., Harmondsworth, UK: Penguin, 1970), 21. Carr bases his discussion on Benedetto Croce's *History as the Story of Liberty* (New York: Meridian Books, 1955).

50. Fabien Locher and Jean-Baptiste Fressoz, "Modernity's Frail Climate: A Climate History of Environmental Reflexivity," *Critical Inquiry* 38, no. 3 (Spring 2012): 579–98.

51. Croce, "The 'History of Nature' and History," in *History: Its Theory and Practice,* 128, 133–34.

52. Ibid., 134.

53. Ibid., 134–35 (emphasis in original).

54. Ibid., 128.

55. Croce, "Nature as History, Not as History Written by Us," in *History as the Story of Liberty,* 290.

56. J. B. S. Haldane, preface to *Everything Has a History* (London: George Allen and Unwin, 1951).

57. Haldane, *Everything,* 11.

58. R. G. Collingwood, *The Idea of History;* R. G. Collingwood, *The Principles of History and Other Writings in Philosophy of History,* ed. and introduced by W. H. Dray and W. J. van der Dussen (1999; repr., Oxford: Oxford University Press, 2003).

59. See James Connelly, Peter Johnson, and Stephen Leach, eds., *R. G. Collingwood: A Research Companion* (2009; repr., London: Bloomsbury Academic, 2015), 83; R. G. Collingwood to Clarendon Press, 9 October 1934, Ref: Clar 46, LB7274, containing comments on J. [B]. S. Haldane, *The Philosophy of a Biologist* (Oxford: Clarendon Press, 1935). This leaves me wondering if Collingwood was a press reader for Haldane's manuscript of this book.

60. Collingwood, *The Idea of History*, 210–11. S. Alexander, "The Historicity of Things," in *Philosophy and History: Essays Presented to Ernst Cassirer*, ed. Raymond Klibansky and H. J. Paton (Oxford: Clarendon Press, 1936), 11–26.

61. Alexander, "Historicity," 12, 15, 16, 21–22.

62. Collingwood, *The Principles,* 56.

63. Ibid., 61.

64. Collingwood, *The Idea*, 212.

65. Ibid., 216.

66. See, for example, C. E. Stamper et al., "The Microbiome of the Built Environment and Human Behavior: Implications for Emotional Health and Well-Being in Postmodern Western Societies," *International Review of Neurobiology* 131 (2016): 289–323, for a study of how modern urbanization impacts the microbiomes and the physical and mental well-being of humans "due in part to decreased exposure to microorganisms that humans have coevolved with."

67. Bruno Latour and Peter Weibel, "Disorientation," in *Critical Zones: The Science and Politics of Landing on Earth*, ed. Bruno Latour and Peter Weibel (Cambridge, MA: MIT Press, 2020), 23. Hartog, too, agrees that the Anthropocene is disorienting but its disorienting quality does not arise simply from the fact that it is not like Christian time, though the fact of that difference is important. See his *Chronos: The West Confronts Time*, trans. S. R. Gilbert (New York: Columbia University Press, 2022; 2020), 226–37.

68. François Hartog, "*Chronos, Kairos,* and the Genesis of Western Time," in "The Eighth *History and Theory* Lecture," *History and Theory* 60, no. 3 (September 2021): 437–38.

69. Benjamin Bratton, "Geoengineering a Rare Earth (for Rare-Earths)," in *The Planet After Geoengineering*, ed. Rania Ghosn et al., (New York: Actar Publishers, 2021), 15.

Notes for Three: Staying with the Present

1. Faisal Devji, "Losing the Present to History," *Modern Intellectual History* (2022): 1–9, https://doi.org/10.1017/S1479244322000117/.

2. See Devji, "Losing the Present to History," 1: "The votaries of a history of the present, for their part, adopt the oldest role in the profession by composing a genealogy or record of contemporary power even if only to hold someone responsible for its evils."

3. Bruno Latour in conversation with Anders Dunkar in Anders Dunkar,

ed., *The Rediscovery of the Earth: 10 Conversations About the Future of Nature* (New York: OR Books, 2020), 16 (emphasis added).

4. François Hartog, *Chronos: l'Occident aux prises avec le Temps* (Paris: Gallimard, 2020), 311–25.

5. Michael Warner, *On the Grid: Climate Change and the Utopia of Green Energy* (New York: Oxford University Press, forthcoming).

6. Mark Williams and Jan Zalasiewicz, *The Cosmic Oasis: The Remarkable Story of Earth's Biosphere* (Oxford: Oxford University Press, 2022), 138.

7. See the discussion in chapter 12 of Pascal Richet, *A Natural History of Time*, trans. John Venerella (1999; repr., Chicago: University of Chicago Press, 2010).

8. Steven Pinker, *Enlightenment Now: The Case for Reason, Science, Humanism, Progress* (New York: Viking, 2018); J. R. McNeill and Peter Engelke, *The Great Acceleration: An Environmental History of the Anthropocene since 1945* (Cambridge, MA: Harvard University Press, 2014).

9. Pinker, *Enlightenment Now*, 39–40.

10. Ibid., 397.

11. Immanuel Kant, *The Conflict of Faculties*, trans. and introduced by Mary J. Gregor (1979; repr., Lincoln: University of Nebraska Press, 1992), 140–71. This particular part of the book was translated by Lewis Beck.

12. Steven Pinker, "Progress," part 2 in *Enlightenment Now*.

13. McNeill and Engelke, *The Great Acceleration*, 41.

14. Jan Zalasiewicz, "The Human Dimension in Geological Time," in *Welcome to the Anthropocene: The Earth in Our Hands*, ed. Nina Möllers et al. (Munich: Deutsches Museum and Rachel Carson Center, 2014), 17.

15. Williams and Zalasiewicz, *The Cosmic Oasis*, 138.

16. Pinker, *Enlightenment Now*, 39.

17. Ibid., 142 (emphasis added).

18. Walter Benjamin, "Franz Kafka: On the Tenth Anniversary of His Death," in *Illuminations*, ed. and intro. Hannah Arendt, trans. Harry Zohn (1973; repr., London: Fontana/Collins, 1982), 133. The odd little creature appears in the short story "Die Sorge des Hausvaters" [The Cares of a Family Man], which was first published in *Ein Landarzt* [A Country Doctor] in 1919.

19. Pinker, *Enlightenment Now*, 53: "A long life is the ultimate blessing." For Pinker's discussion on climate engineering, see ibid., 153–54.

20. Pinker, *Enlightenment Now*, 154–55 (emphasis added).

21. Ibid., 153.

22. Williams and Zalasiewicz, *The Cosmic Oasis*, 151.

23. Christine Winter, "Introduction: What's the Value of Multispecies Justice?," *Environmental Politics* 31, no. 2 (2022): 255, 256.

24. Frédéric Neyrat, *The Unconstructable Earth: An Ecology of Separable* (New York: Fordham University Press, 2019), 15.

25. Williams and Zalasiewicz, *The Cosmic Oasis*, 154.

26. Ibid., 116. See also 148, 149.

27. Ibid., 141–42.

28. Gianmaria Colpani, Jamila M. H. Mascat, and Katrine Smiet, "Postcolonial Responses to Decolonial Interventions," *Postcolonial Studies* 25, no. 1 (2022): 3. See also Gianmaria Colpani, "Crossfire: Postcolonial Theory between Marxist and Decolonial Critiques," in the same issue, 54–72.

29. Déborah Danowski and Eduardo Viveiros de Castro, *The Ends of the World*, trans. Rodrigo Nunes (Cambridge: Polity, 2017. First published in Portuguese, 2014).

30. See Dipesh Chakrabarty, *The Climate of History in a Planetary Age* (Chicago: University of Chicago Press, 2021), chapter 1.

31. Danowski and de Castro, *Ends of the World*, 90 (emphasis added).

32. Ibid.

33. Eduardo Viveiros de Castro, *Cannibal Metaphysics*, ed. and trans. Peter Skafish (Minneapolis: Univocal Publishing, 2014; first published in French, 2009).

34. Ibid., 48 (emphasis added); see also 40: "Anthropology is ready to fully assume its new mission of being the theory/practice of the permanent decolonization of thought."

35. Danowski and de Castro, *Ends of the World*, 90. See de Castro, *Cannibal Metaphysics*, chapters 2 and 3, for explication of the ideas of "perspectivism" and "multinaturalism."

36. De Castro, *Cannibal Metaphysics*, 97.

37. De Castro, *Cannibal Metaphysics*, 108. Bruno Latour, *We Have Never Been Modern*, trans. Catherine Porter (Cambridge, MA: Harvard University Press, 1993; first published in French, 1991). In personal conversations, Latour has mentioned to me the influence of Deleuze on his thinking.

38. De Castro, *Cannibal Metaphysics*, 108–9.

39. Ibid., 110.

40. See Gilles Deleuze and Félix Guattari, *Anti-Oedipus: Capitalism and Schizophrenia*, trans. Robert Hurley, Mark Seem, and Helen Lane (Minneapolis: University of Minnesota Press, 1983; first published in French, 1972), chapter 3, 139–53, 217–22.

41. De Castro, *Cannibal Metaphysics*, 109.

42. Ibid., 109 n. 64.

43. Michel Foucault, "Preface," to Deleuze and Guattari, *Anti-Oedipus*, xiii.

44. Deleuze and Guattari, *Anti-Oedipus*, 139.

45. Ibid., 195.

46. Ibid., 140.

47. The reference here is to Friedrich Engels, *The Origin of the Family* (New York: International Publishers, 1942).

48. Deleuze and Guattari, *Anti-Oedipus*, 145–46.

49. Gilles Deleuze and Félix Guattari, *A Thousand Plateaus: Capitalism and Schizophrenia*, trans. Brian Massumi (1987; repr. Minneapolis: University of Minnesota Press, 1991; first published in French, 1980).

50. Deleuze and Guattari, *A Thousand Plateaus*, 357.

51. Ibid., 359.

52. Ibid., 360.

53. Ibid.

54. Bruno Latour, *Facing Gaia: Eight Lectures on the New Climatic Regime*, trans. Catherine Porter (Cambridge: Polity, 2017; first published in French, 2015).

55. Latour, *Facing Gaia*, 251.

56. Danowski and de Castro, *Ends of the World*, 92–93 (emphasis added).

57. Ibid., 91.

58. Ibid., 96.

59. Ibid., 91.

60. Ibid.

61. Ibid., 90.

62. Ibid., 94–95.

63. Ibid., 95 (emphasis added).

64. Here I refer to the work of Danielle Celermajer, Christine Winter, Jeremy Bendik-Keymer, Julia Gibson, and others that I have, much to my shame, only recently encountered.

65. On de Castro's permanent decolonization, see n. 34, above. B. R. Ambedkar, *The Annihilation of Caste* (Jalandhar: Bheem Patrika Publications, n.d.), 92, 129, 131. See also the discussion in chapter 2 of this book.

66. On the genocidal logic of settler-colonial practices, see Patrick Wolfe, *Traces of History: Elementary Structures of Race* (London: Verso, 2016).

67. Michel Foucault cited in Naoki Sakai, *The End of Pax Americana: The Loss of Empire and Hikikomori Nationalism* (Durham, NC: Duke University Press, 2022), 177.

68. Danowski and de Castro, *Ends of the World*, 95.

69. See Jonathan Lear, *Radical Hope: Ethics in the Face of Cultural Devastation* (2006; repr., Cambridge, MA: Harvard University Press, 2008), chapter 3.

70. Anna Lowenhaupt Tsing, *The Mushroom at the End of the World: On the Possibility of Life in Capitalist Ruins* (Princeton: Princeton University Press, 2015), 305, n.12.

71. Eric Dean Wilson, *After Cooling: On Freon, Global Warming, and the Terrible Cost of Comfort* (New York: Simon and Schuster, 2021).

72. Wilson, *After Cooling*, 338.

73. Ibid., 336.

74. Ibid., 336–37 (emphasis added).

75. Ibid., 338 (emphasis added).

76. Sunita Narain and Anil Agarwal, *Global Warming in an Unequal World: A Case of Environmental Colonialism* (New Delhi: Centre for Science and Environment, 1991).

77. Wilson, *After Cooling*, 17; Kathryn Yusoff, *A Billion Black Anthropocenes or None* (Minneapolis: University of Minnesota Press, 2018).

78. Wilson, *After Cooling*, 17.

79. Ibid., 17; Yusoff, *Black Anthropocenes*, preface, xiii (emphasis added).

80. Jacques Derrida, "Force of Law: The 'Mystical Foundation of Authority,'" in *Deconstruction and the Possibility of Justice*, ed. Drucilla Cornell, Michel Rosenfeld, David Gray Carlson (London: Routledge, 1992), 4.

81. See Adom Getachew, *Worldmaking After Empire: The Rise and Fall of Self-Determination* (Princeton: Princeton University Press, 2019); Ashis Nandy, *The Intimate Enemy: Loss and Recovery of Self after Colonialism* (New Delhi: Oxford University Press, 1983); Dipesh Chakrabarty, *The Crises of Civilization: Exploring Global and Planetary Histories* (New Delhi: Oxford University Press, 2018).

82. Yusoff, *Black Anthropocenes*, 21 (emphasis in original).

83. Zoe Todd, "An Indigenous Feminist's Take on the Ontological Turn: 'Ontology' is Just Another Word for Colonialism," *Journal of Historical Sociology* 29, no. 1 (March 2016): 14.

84. Winter, "Value of Multispecies Justice," 255.

85. See the discussion in Ajantha Subramaniam and Paula Chakravarti, "Why is Caste Inequality Still Legal in America?," *New York Times*, May 25, 2021, accessed July 30, 2022, www.nytimes.com/2021/05/25/opinion/caste -discrimination-us-federal-protection.html/.

86. Williams and Zalasiewicz, *The Cosmic Oasis*.

87. Tsing, *The Mushroom*, vii.

88. See the discussion in Seth Epstein, "Rights of Nature, Human Species Identity, and Political Thought in the Anthropocene," *The Anthropocene Review* (2 May 2022): https://doi.org/10.1177/20530196221078929.

89. Aase J. Kvanneid, "Climate Change, Gender, and Rural Development: Making Sense of Coping Strategies in the Shivalik Hills," *Contributions to Indian Sociology* 55, no. 3 (2021): 407.

90. See Bruno Latour and Dipesh Chakrabarty, "Conflicts of Planetary Proportions," *Journal of Philosophy of History* 14, no. 3 (2020); Christophe Bonneuil, "Der Historiker und der Planet—Planetaritätsregimes an der Schnittstelle von Welt-Ökologien, ökologischen Reflexivitäten und Geo-Mächten," in *Gessellschaftstheorie im Anthropozän*, ed. Frank Adloff and Sighard Neckel (Frankfurt: Campus Verlag, 2020), 55–94.

91. Bonneuil, "Der Historiker," 73–74.

92. Christian Schwägerl, "'We Aren't Doomed': An Interview with Paul Crutzen," in Möllers, *Welcome to the Anthropocene*, 36.

93. I am thinking in particular of the Indian paleobotanist Birbal Sahni (1891–1949). On his collaboration with the Chinese scientist Hsü Jen (1910–1992) in search of an "Asian" paleobotany, see Arunabh Ghosh, "Trans-Himalayan Science in Mid-Twentieth Century China and India: Birbal Sahni, Hsü Jen, and a Pan-Asian Paleobotany," *International Journal of Asian Studies* 19, no. 3 (May 2021): 1–23, https://doi:10.1017/S1479591421000292. See also Ashok Sahni, "Birbal Sahni and His Father Ruchiram Sahni: Science in Punjab Emerging from the Shadows of the Raj," *Indian Journal of History of Science* 54, no. 3 (2018): T160–T166, https://doi 10.16943/ijhs/2018/v53i4/49539/. For Cohn's luminous critique of colonial knowledge, see Bernard S. Cohn,

Colonialism and Its Forms of Knowledge (Princeton: Princeton University Press, 1996).

94. For the case of an Indian historian inspired by Ranke, see my book *The Calling of History: Sir Jadunath Sarkar and His Empire of Truth* (Chicago: University of Chicago Press, 2015).

95. Homi Bhabha's critical discussion of "mimicry" in his *The Location of Culture* (London: Routledge, 1994) still remains relevant here.

96. Achille Mbembe, "Proximity without Reciprocity," in *Out of the Dark Night: Essays on Decolonization* (New York: Columbia University Press, 2021), 101: "There is a kind of 'we' that . . . takes form on a global scale and especially in the act by which one shares differences."

97. Tsing, *The Mushroom*, 158–59.

98. Robin Wall Kimmerer, *Braiding Sweetgrass: Indigenous Wisdom, Scientific Knowledge, and the Teaching of Plants* (Minneapolis: Milkweed Editions, 2013).

99. Kimmerer, *Braiding Sweetgrass,* 48.

100. Ibid., 48–49.

101. Ibid., 49.

102. Ibid., 49.

103. Homi K. Bhabha, introduction to *The Location of Culture* (London: Routledge, 1994), 13.

104. Karl Marx, *Economic and Philosophical Manuscripts of 1844,* trans. unknown (1932; repr. Moscow: Progress Publishers, 1958); M. K. Gandhi, *Hind Swaraj* (1909) in Gandhi, *Hind Swaraj and Other Writings*, ed. Anthony Parel (Cambridge: Cambridge University Press, 2009).

About the Author

DIPESH CHAKRABARTY is the Lawrence A. Kimpton Distinguished Service Professor in History, South Asian Languages and Civilizations, and the College at the University of Chicago. He is an associate in the Faculty of Arts and Social Sciences at the University of Technology in Sydney, Australia, and holds a distinguished visiting professorship at the Australian National University. Chakrabarty is the recipient of the 2014 Toynbee Foundation Prize for his contributions to global history and of the 2019 Tagore Memorial Prize, awarded by the Government of West Bengal, for his book *The Crises of Civilization*. He was elected an honorary fellow of the Australian Academy of the Humanities in 2006 and a fellow of the American Academy of Arts and Sciences in 2004.